ADULTS OF THE BRITISH
AQUATIC HEMIPTERA HETEROPTERA:
A KEY WITH ECOLOGICAL NOTES

AUTHOR'S PREFACE

This key is dedicated to the memory of Dr T. T. Macan, who stimulated my interest in water bugs some thirty-five years ago when I attended Easter Classes held at The Ferry House. That stimulus has borne fruit in this new key, which owes much to Dr Macan's own research and previous keys to water bugs. I also take the opportunity to record my gratitude and appreciation of the kind help and encouragement that I received from Dr Macan for many years until his death in 1985.

Crewe & Alsager College of Higher Education   Alan A. Savage
December 1988

PLYMOUTH POLYTECHNIC LIBRARY

Accn No   212892-8

Class No   595. 754   SAV

Contl No   0900386487

Published by the Freshwater Biological Association, The Ferry House, Ambleside, Cumbria LA22 oLP
© Freshwater Biological Association 1989

ISBN 0 900386 48 7
ISSN 0367-1887

PREFACE

The first title in the Association's series of Scientific Publications appeared in 1939 and was a key to British Corixidae by T. T. Macan. A later key, dealing with the other water bugs, was also written by Dr Macan and appeared in 1941 as Scientific Publication number 4. Both keys were combined in Scientific Publication number 16 which appeared in 1956 with a second, revised edition in 1965. It is therefore fitting that the 50th publication in the Association's series should appear 50 years after the first and be a new key to the British water bugs.

The Association is indeed fortunate that Dr Alan Savage agreed to be the author of the new publication. He has worked on water bugs for many years and his wealth of knowledge is apparent in this text. As in all recent publications in this series, there is a large section on the ecology of the water bugs. This review should serve as a useful introduction to the literature for those who wish to pursue ecological studies on water bugs. I hope that all will welcome this publication in the Association's series.

The Ferry House   J. G. Jones
January 1989   *Director*

CONTENTS

INTRODUCTION

The key in the present publication is a revised version of its immediate predecessor in the series (Macan 1965a). This work itself was a development of earlier keys by Macan (1939, 1941, 1956). Experience of using Macan (1965a) with some thirty cohorts of students in tertiary education and teachers in both primary and secondary schools suggests that its fundamental structure is satisfactory. However, this experience has resulted in many detailed changes in the text and considerable rearrangement of, and additions to, the illustrations.

The main change in the key itself is the inclusion of a newly described species, *Corixa iberica* Jansson, 1981. It was initially described from specimens from the South West Iberian Peninsula but was later noted in British material (Jansson 1986). The specimens noted by Jansson have been examined together with thirty–seven further specimens discovered in existing collections. British material has been used in the construction of this key. It is also noteworthy that *Limnoporus rufoscutellatus* (Latreille), already included in the key, now may be regarded as a member of the indigenous fauna (O'Connor 1986).

A major change in the work as a whole is the inclusion of a substantial section on the ecology of water bugs. There have been important developments in the study of these insects in the past two decades, particularly in North America and Scandinavia, which have made major contributions to our knowledge of the group. The inclusion of recent work has necessitated the exclusion of detailed reference to much of the earlier work but it may be traced from the references given. In particular, Southwood & Leston (1959) is a valuable national source while the accounts of semi-aquatic bugs (Andersen 1982) and Corixidae (Jansson 1986) provide a broader perspective.

GENERAL STRUCTURE AND MORPHOLOGICAL TERMS

The general structure of the aquatic Hemiptera Heteroptera differs markedly in the various families. Thus, it is not possible to give a single text-figure illustrating all of the diagnostic features used in the key. A species from each family is illustrated in the key to families (figs 1–12) and the major diagnostic features used in the identification of species of Gerridae, Notonectidae and Corixidae are shown in figs 20, 28, 34 respectively. These, together with figs 1–12, should be consulted for other families.

The positions of structures are normally described using the following terms: dorsal (above), ventral (below), lateral (side), anterior (front), posterior (behind), proximal (nearer to body centre) and distal (further from body centre). On occasion, it has seemed too clumsy or over-pedantic to maintain this system and plain language has been used when it is sufficiently precise. The numbering of parts of structures begins at their proximal ends.

Head. The *antennae* are long and conspicuous in the surface dwellers and amphibious taxa but inconspicuous in the sub-aquatic taxa (e.g. figs 1, 7, 8, 12). *Ocelli* (small eyespots on the dorsal region posterior to the eyes) may be present (fig. 4H) or absent (fig. 4M). The *mouthparts* of Hemiptera are typically formed into a tube-like, piercing and sucking *rostrum* or *beak* (e.g. fig. 20r). However, it is much reduced and inconspicuous in Corixidae.

Thorax. There are three segments in the thorax: the anterior *prothorax* with the anterior (fore) pair of legs; the *mesothorax* with the *hemielytra* (when present) and middle pair of legs; the posterior *metathorax* with wings (when present) and posterior (hind) pair of legs (e.g. figs 19, 20). The dorsal plates of each of these segments are the *pronotum*, *mesonotum* and *metanotum* respectively. In addition there is a posteriorly pointing, triangular plate in the posterior mid-line of the mesonotum proper, the *scutellum* (figs 28sc, 66sc), which is visible in some taxa. The pronotum has a longitudinal *keel* in Gerridae and some Corixidae (fig. 20k). Similarly, the ventral plates are the *prosternum*, *mesosternum* and *metasternum*. The *metasternal xiphus* is a diagnostic feature in Corixidae (fig. 43).

Each *hemielytron* consists of a mid-proximal *clavus*, a lateral-proximal *corium* and a distal *membrane* (figs 28, 34). The *hemielytra* and *wings* may be absent, of intermediate lengths, or fully developed; many species show *wing polymorphism*. Each *leg*, typically, has six parts: a proximal short *coxa*, an inconspicuous *trochanter*, a conspicuous *femur*, *tibia*, *tarsus* (which may be subdivided) and one or two *claws* (fig. 34). The *pala*, a flattened anterior tarsus, occurs in most Corixidae (fig. 37).

Abdomen. Each abdominal segment consists of a dorsal plate, the *tergum* (*tergite*) and a ventral plate, the *sternum* (*sternite*). Sometimes, where the two meet there are lateral folds, the *connexiva* (fig. 20). The fifth tergum of male Micronectinae bears a *prestridular flap* (fig. 66*ps*). Males of most Corixinae bear a black, toothed plate, the *strigil*, on the posterior margin of the sixth tergum; it is on the right in some species and on the left in others (figs 34*s*, 35*s*, 59*gr a*, *cr a*, *pp a*). The *male genitalia* provide diagnostic features in Veliidae, Gerridae, Notonectidae and Corixidae. The ninth segment is more or less modified to form a *genital capsule* (figs 28, 31, 35): to it are attached a central *aedeagus*, which may contain chitinous *sclerites*, and the lateral *parameres* (e.g. figs 15, 22, 35, 41).

CLASSIFICATION AND CHECK-LIST

Classification and nomenclature are essentially based on Kloet & Hincks (1964) as this remains the standard reference work for the British Isles. However, a number of changes have been made, some absolutely necessary and some a matter of opinion, in the light of more recent studies. The aim has been accuracy, simplicity, consistency and agreement with recent work, particularly as exemplified by lists in Maitland (1977), Nieser (1978), Andersen (1982), Jansson (1986) and Štys & Jansson (1988).

Classification above the level of genus has been simplified by the omission of superfamilies, and the terms Gerromorpha and Nepomorpha have been used in place of Amphibicorisae and Hydrocorisae respectively.

The family Corixidae is divided simply into three subfamilies, namely, Micronectinae, Cymatiainae and Corixinae, following Jansson (1986). The division of Corixinae into the tribes Glaenocorisini and Corixini is rejected and thus the subfamily proceeds directly to genera. The change in spelling to Cymatiainae rather than Cymatiinae is because the latter is a homonym of Cymatiidae, Iredale, Mollusca: Gastropoda (Jansson 1986).

The genera *Gerris* and *Sigara* are treated in a broad sense but the divisions into further genera are included for convenience.

The most notable change at species level is the addition of *Corixa iberica* Jansson. A few synonyms are included if they are likely to be found in museum collections or are common in the literature. A comprehensive list of synonyms for species of Corixidae may be found in Jansson (1986).

The treatment of subspecies presents a number of potential problems. Where the British subspecies is the type, no reference is made to subspecies in the check-list. For instance, *Sigara concinna concinna* (Fieber) appears simply as *S. concinna* (Fieber) since *S. concinna amurensis* (Jaczewski) occurs only in Asia. In contrast, where the British species is not the type, or where more than one subspecies occurs as in *Glaenocorisa propinqua* (Fieber), then subspecific names are included in both check-list and key, thus forming trinomials.

One new trinomial is of particular interest. Kanyukova (1986) proposed that *Microvelia umbricola* Wróblewski is a synonym of *M. buenoi* Drake. I have compared specimens of the two original species, but only *M. umbricola* was available in quantity. They are certainly closely similar. However, the hemielytral patterns of winged specimens are different in the two species. Furthermore, Drake (1920) states that the posterior borders of the pronotum and mesonotum of the wingless forms of *M. buenoi* are smoothly rounded whereas they are sinuate in British *M. umbricola*. This distinction is quite clear in males, less so in females. I propose that the synonomy should be accepted but the North American and European populations should be given separate subspecific status. Thus, the North American type is *Microvelia buenoi buenoi* Drake while the European subspecies is *Microvelia buenoi umbricola* Wróblewski. It is possible that Asian specimens seen by Kanyukova may be intermediate in structure and status.

Two widely known infraspecific names are rejected and regarded as synonyms. The subspecies *Cymatia coleoptrata insularis* Walton is regarded as a synonym of *Cymatia coleoptrata* (Fabricius). It appears that there is now some doubt concerning the origin of the putative Scottish specimens (Walton, personal communication) and the parameres, which were used as a criterion for subspecific status (Walton 1942), are very variable in this species. Indeed, no authentic Scottish specimens have been found. My own experience suggests that the North West Midlands (MD, Table 3) are close to the northern limit of *C. coleoptrata*, since it occurs there frequently in some habitats for a number of years and then becomes very scarce for a time after a hard winter. Similarly, the subspecies or variety *Callicorixa wollastoni caledonica* (Kirkaldy) is regarded as a synonym of *C. wollastoni* (Douglas & Scott). More than 1,000 specimens have been examined during recent years. Many intermediates were seen between the typical form which shows little contrast between light and dark areas and, for instance, specimens from Loch Leven which show a clear contrast. The degree of infuscation of the light parts of Corixidae is associated with habitat type (Popham 1943). I have found that *Callicorixa praeusta* (Fieber) and *Corixa punctata* (Illiger) show less contrast between light and dark areas when they occur in dark peaty habitats. *C. wollastoni* is typical of such places and only deviates from the normal form when found in unusual habitats, where it appears to be an example of purely phenotypic variation.

Finally, two further possible problems may be mentioned. *Notonecta obliqua* Gallen possesses a variety, *delcourti* Poisson, which is readily recognisable. It appears unlikely that it has any taxonomic status but is an example of minor genotypic variation which may be of value in comparative population studies. Therefore it is omitted from the check-list but retained in the key.

In two closely related species, *Sigara striata* (Linnaeus) and *S. dorsalis* (Leach), intermediate forms occur in South East England, (SE, Table 3; figs 52, 53, 71). However, in hybridisation experiments, Jansson (1979c) has clearly demonstrated that considerable genotypic differences exist between these species. It seems likely that they are an example of incipient post-glacial geographical speciation. Similar differences occur in isolated populations of *Arctocorisa carinata* (Sahlberg) and *A. germari* (Fieber) (Jansson 1979b, 1980).

It should be noted that British specimens described as *S. striata*, prior to 1955, are almost certainly *S. dorsalis*. There is further potential confusion since the name *S. lacustris* (Macan), a synonym of *S. dorsalis*, was used for a short period subsequent to 1955 (Macan 1954b, 1955b; Macan & Leston 1978).

TABLE 1. A CHECK-LIST OF THE BRITISH AQUATIC HEMIPTERA HETEROPTERA
(Genera and subgenera in parentheses are not used in this key. sg.=Subgenus;
ss.=Subspecies; f.=Form.)

FAMILY SUBFAMILY	GENUS	*Species* = *synonym*	

INFRAORDER GERROMORPHA

MESOVELIIDAE	MESOVELIA Mulsant & Rey, 1852	*furcata* Mulsant & Rey, 1852	(1)
HEBRIDAE	HEBRUS Curtis, 1833*	*pusillus* (Fallén, 1807)	(2)
		ruficeps (Thomson, 1871)	(3)
HYDROMETRIDAE	HYDROMETRA Latreille, 1796	*gracilenta* Horvath, 1899	(4)
		stagnorum (Linnaeus, 1758)	(5)
VELIIDAE			
VELIINAE	VELIA Latreille, 1804	*caprai* Tamanini, 1947	(6)
		= *currens*: auctt., nec (Fabricius, 1794)	
		saulii Tamanini, 1947	(7)
MICROVELIINAE	MICROVELIA Westwood, 1834	*pygmaea* (Dufour, 1833)	(8)
		reticulata (Burmeister, 1835)	(9)
		= *schneideri* (Scholtz, 1847)	
		= *pygmaea*: Douglas & Scott, 1865 Saunders, 1892 nec (Dufour, 1833)	
		buenoi Drake, 1920	(10)
		ss. *umbricola* Wróblewski, 1938	
GERRIDAE			
GERRINAE	GERRIS Fabricius, 1794	*costai* (Herrich-Schäffer, 1853)	(11)
		ss. *poissoni* Wagner & Zimmerman, 1955	
		lateralis Schummel, 1832	(12)
		ss. *asper* (Fieber, 1861)	
		thoracicus Schummel, 1832	(13)
		gibbifer Schummel, 1832	(14)
		argentatus Schummel, 1832	(15)
		lacustris (Linnaeus, 1758)	(16)
		odontogaster (Zetterstedt, 1828)	(17)
	(AQUARIUS Schellenberg, 1800)	*najas* (DeGeer, 1773)	(18)
		paludum (Fabricius, 1794)	(19)
	LIMNOPORUS Stål, 1868	*rufoscutellatus* (Latreille, 1807)	(20)

*Synonym NAEOGEUS, Laporte, 1833   ()

INFRAORDER NEPOMORPHA

NEPIDAE

NEPINAE	NEPA Linnaeus, 1758	*cinerea* Linnaeus, 1758	(21)
RANATRINAE	RANATRA Fabricius, 1790	*linearis* (Linnaeus, 1758)	(22)
NAUCORIDAE	ILYOCORIS Stål, 1861	*cimicoides* (Linnaeus, 1758)	(23)
APHELOCHEIRIDAE	APHELOCHEIRUS Westwood, 1833	*aestivalis* (Fabricius, 1794)	(24)

f. *montandoni* Horvath, 1899

NOTONECTIDAE NOTONECTA Linnaeus, 1758

glauca Linnaeus, 1758 (25)
marmorea Fabricius, 1803 (26)
ss. *viridis* Delcourt, 1909
= *halophila* Edwards, 1918
obliqua Gallen, 1787 (27)
= *furcata* Fabricius, 1794
maculata Fabricius, 1794 (28)

PLEIDAE PLEA Leach, 1817

leachi McGregor & Kirkaldy, 1899 (29)
= *atomaria* (Pallas, 1771)
= *minutissima* (Füssly, 1775), nec
(Linnaeus, 1758)

CORIXIDAE

MICRONECTINAE MICRONECTA Kirkaldy, 1897

scholtzi (Fieber, 1847) (30)
= *meridionalis* (Costa, 1863)
minutissima (Linnaeus, 1758) (31)
poweri (Douglas & Scott, 1869) (32)
= *minutissima*: auctt. Brit., nec
(Linnaeus, 1758)

CYMATIAINAE CYMATIA Flor, 1860

bonsdorffii (Sahlberg, C., 1819) (33)
coleoptrata (Fabricius, 1776) (34)
= ss. *insularis* Walton, 1942

CORIXINAE GLAENOCORISA Thomson, 1869

propinqua (Fieber, 1860) (35)
ss. *propinqua* (Fieber, 1860)
ss. *cavifrons* (Thomson, 1869)
= *alpestris* (Douglas & Scott, 1870)
= *quadrata* Whalley, 1930

CALLICORIXA White, F. B., 1873

praeusta (Fieber, 1848) (36)
= *wollastoni*: (Saunders, 1892)
nec (Douglas & Scott, 1865)
wollastoni (Douglas & Scott, 1865) (37)
= *cognata*: (Douglas & Scott, 1870)
nec (Fieber, 1860)
= *caledonica* (Kirkaldy, 1897)

CORIXA Geoffroy, 1762

dentipes (Thomson, 1869) (38)
punctata (Illiger, 1807) (39)
= *geoffroyi* Leach, 1817
iberica Jansson, 1981 (40)
affinis Leach, 1817 (41)
panzeri (Fieber, 1848) (42)

(*Continued on p. 12*)

CORIXINAE

HESPEROCORIXA Kirkaldy, 1908   *linnaei* (Fieber, 1848)   (43)
       sahlbergi (Fieber, 1848)   (44)
       castanea (Thomson, 1869)   (45)
       moesta (Fieber, 1848)   (46)

ARCTOCORISA Wallengren, 1894   *carinata* (Sahlberg, C., 1819)   (47)
       germari (Fieber, 1848)   (48)

SIGARA Fabricius, 1775   *dorsalis* (Leach, 1817)   (49)
       = *striata*: auctt. Brit. nec
         (Linnaeus, 1758)
       = *lacustris* (Macan, 1954)
       striata (Linnaeus, 1758)   (50)

(sg. SUBSIGARA Stichel, 1935)   *distincta* (Fieber, 1848)   (51)
       = *douglasi* (Fieber, 1865)
       falleni (Fieber, 1848)   (52)
       fallenoidea (Hungerford, 1926)   (53)
       = *pearcei* Walton, 1936
       fossarum (Leach, 1817)   (54)
       scotti (Douglas & Scott, 1868)   (55)

(sg. VERMICORIXA Walton, 1940)   *lateralis* (Leach, 1817)   (56)
       = *hieroglyphica* (Dufour, 1833)

(sg. PSEUDOVERMICORIXA   *nigrolineata* (Fieber, 1848)   (57)
Jaczewski, 1962)

(PARACORIXA Poisson, 1957)   *concinna* (Fieber, 1848)   (58)

(sg. RETROCORIXA Walton, 1940)   *limitata* (Fieber, 1848)   (59)
       semistriata (Fieber, 1848)   (60)
       venusta (Douglas & Scott, 1869)   (61)

(sg. HALICORIXA Walton, 1940)   *selecta* (Fieber, 1848)   (62)
       stagnalis (Leach, 1817)   (63)
       = *lugubris* (Fieber, 1848)

COLLECTION AND PRESERVATION

The qualitative collection of water bugs is a simple process but requires modification according to the niches of the taxa required. Fully aquatic bugs swimming in the water may be collected by vigorous sweeping just above the substratum with a long-handled pond net. *Aphelocheirus aestivalis*, which occurs in stony streams, is most easily collected by disturbing stones with the feet while sweeping vigorously immediately *upstream*. Surface dwellers may be collected by sweeping with the pond net only slightly submerged; Gerridae may be stalked individually. Hebridae occur at the water margin among *Sphagnum* spp.. Clumps of the plants should be gathered and placed in a dish of water; the bugs then rise to the surface. Information may be gained on the particular niches of the different species by restricting collections to particular sets of environmental conditions, i.e. biotopes. For example, separate net sweeps should be made in open water, on the edge of vegetation, and amongst vegetation at varying distances from the edge and in different species of plants.

Collections which attempt both qualitative and quantitative estimates of communities of these lively organisms present certain problems. The simplest method is to make a standard net sweep over a given area of substratum. For instance, Savage (1979a, 1981) collected Corixidae using a robust hand net of rectangular cross section having an area of $500\,cm^2$ ($25\,cm$ wide $\times 20\,cm$ high) and a mesh of 7.5 strands cm^{-2}. Two imaginary points A and B, $2\,m$ apart, were selected. The net was placed in the water at A and moved rapidly just above the substratum to B, back to A and then to B again with the opening always facing the direction of movement. Thus, an area of approximately $0.5\,m^2$ was covered. Ten such sweeps were made in a particular biotope, the numbers of each species counted and means and confidence limits calculated (Elliott 1977). Thus, numerical estimates of populations per unit area were made. Crisp (1962a, b) used a slightly different version, adapted for deeper water, to estimate numbers of *Arctocorisa germari* in an upland reservoir. Indeed, the details of the method must be adapted to a particular habitat and perfected in the most 'difficult' biotope, such as dense reed beds, when comparative studies are made. Popham (1964) counted Corixidae in a given area by placing metal squares (quadrats) on the substratum while Henrikson & Oscarson (1978a) and Behr (1988) have designed traps for aquatic insects which visit the surface. The numbers of Hebridae may be estimated by removing a known volume of *Sphagnum* spp. with a sharp metal corer prior to extraction and counting. Thoroughly satisfactory methods for all taxa of water bugs have yet to be devised and there is a need for further study and experiment. A valuable starting point for the development of quantitative methods for collecting benthic macroinvertebrates, is provided by the annotated-bibliographies of Elliott & Tullett (1978, 1983).

Specimens may be taken back to the laboratory in tubes or polyethylene bags containing a little water and plenty of air. All taxa may be maintained for short periods in aquaria containing water a few cm deep, material from the substratum, stones and plants, the whole being covered with gauze.

Specimens should be killed with ethyl acetate vapour. A series of glass tubes approximately $10 \, cm \times 2.5 \, cm$ are convenient. A layer of celluloid chips a few mm deep is placed in each tube and a few ml of ethyl acetate added. A plug of cotton wool about 1 cm long is inserted and covered by a folded cone of filter paper to prevent the specimens from becoming entangled. The tubes should be corked so that the slow escape of vapour prevents condensation and consequent damage to specimens. The tubes are effective for a few weeks and may be reactivated by the addition of ethyl acetate directly onto the celluloid chips. Specimens left in the tubes remain relaxed.

Specimens may be preserved in an alcohol (ethanol)-based mixture, pinned or mounted on card. A suitable mixture comprises 80 parts by volume 70% ethanol, 10 parts 40% formalin and 10 parts glycerol. When specimens are pinned or mounted care must be taken, by reference to the diagnostic features used in the key, to ensure that the necessary structures are visible. Water soluble glues, such as Gum Tragacanth, must be used for mounting. Data labels must include date of capture, National Grid Reference and a note of the biotope.

EXAMINATION

Methods for the examination of particular diagnostic features are given in the key. Nevertheless, a few general points may be of assistance to those unfamiliar with the aquatic Hemiptera Heteroptera.

Specimens should be surface dry, relaxed and unmounted. A good quality low power binocular microscope with magnifications of approximately ×5, ×10 and ×40 is ideal for basic identification. The most important requisite is a good light source placed to one side; fibre optics systems with flexible stems and using halogen bulbs are suitable. Confirmation of an identification frequently requires the dissection and removal of small parts. The detailed instructions given in the relevant part of the key should be noted and followed by examination on a microscope slide at magnification ×100.

The absolute or relative lengths of specimens and structures are used on occasion. It is wise to measure accurately rather than guess! A calibrated eyepiece graticule, a small pair of dividers, or millimetre squared graph paper may be used, as appropriate.

KEY TO FAMILIES

Adults of British aquatic Hemiptera Heteroptera may be separated from other aquatic insects by the following characters: hemielytra, when present, divided into distinct regions (figs 28, 34); mouthparts formed into a long, pointed, piercing, sucking rostrum (fig. 20) or, as in Corixidae, inconspicuous; biting mouthparts never present.

When a key couplet is not reached directly from the preceding one, the number of the couplet from which it originates is given in parentheses.

The range of body lengths is given for each family. They are seldom exclusive but should prove helpful.

1   Antennae conspicuous, much longer than the head. Animals dwelling on the surface of the water in the open or among vegetation (figs 1, 3–7)—   **2**

—   Antennae, which are not seen in dorsal view, much shorter than the head and concealed in pits on the ventral side. Animals dwelling in the water (figs 8–12)—   **6**

2   Head many times longer than wide; eyes situated some distance from the anterior margin of the pronotum (fig. 1).
Length 7·5–12·0 mm—   HYDROMETRIDAE, p. 30

—   Head little, if at all, longer than wide; eyes close to anterior margin of pronotum—   **3**

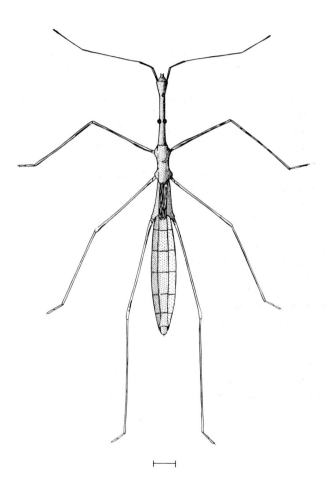

Fig. 1. Hydrometridae (*Hydrometra stagnorum*). (Scale line 1 mm).

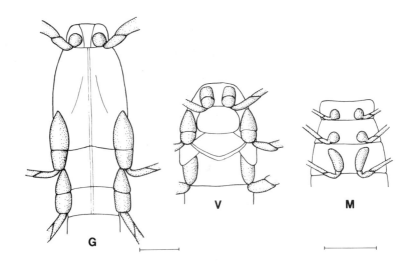

Fig. 2. Ventral view of the thorax showing the insertion of the legs in: *G*, Gerridae (*Gerris* sp.); *V*, Veliidae (*Velia* sp.); *M*, Mesoveliidae (*Mesovelia furcata*). (Scale lines 1 mm).

3   All legs inserted towards the centre line of the thorax (fig. 2M). Length 3·0–3·5 mm—   MESOVELIIDAE, p. 29

—   Legs, particularly the posterior legs, inserted towards or at the lateral margins of the thorax (fig. 2G, V)—   4

4   Antennae five-segmented, the distal three segments thinner than the proximal two; ocelli present (fig. 4H). Length 1·0–2·0 mm—   HEBRIDAE, p. 29

—   Antennae four-segmented, all segments roughly equal in thickness; ocelli absent—   5

Occasionally, specimens of Hebridae may be collected with semi-terrestrial bugs of the family Dipsocoridae (Leptopodidae), which are of similar size and also occur amongst *Sphagnum* spp.. Dipsocoridae possess only four antennal segments, the proximal two being short and thick whereas the distal two are long and filamentous.

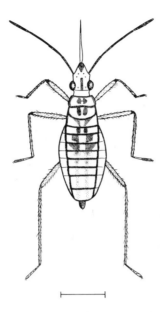

Fig. 3. Mesoveliidae (*Mesovelia furcata*). (Scale line 1 mm).

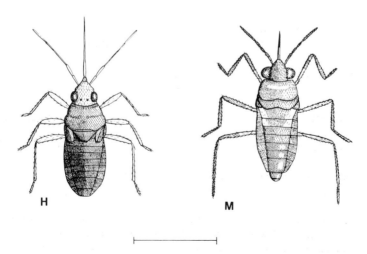

Fig. 4. *H*, Hebridae (*Hebrus ruficeps*); *M*, Veliidae (*Microvelia buenoi umbricola*). (Scale line 1 mm).

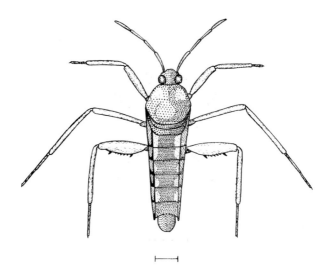

Fig. 5. Veliidae (*Velia caprai*). (Scale line 1 mm).

5   Middle legs inserted roughly midway between the anterior and pos-
     terior legs (fig. 2V); posterior femora not extending beyond tip of
     abdomen (figs 4M, 5, 6V).
     Length 1·4–8·0 mm—   VELIIDAE, p. 32

—   Middle legs inserted much nearer the posterior than the anterior legs
     (fig. 2G); posterior femora extending well beyond tip of abdomen
     (figs 6G, 7).
     Length 6·5–18·0 mm—   GERRIDAE, p. 42

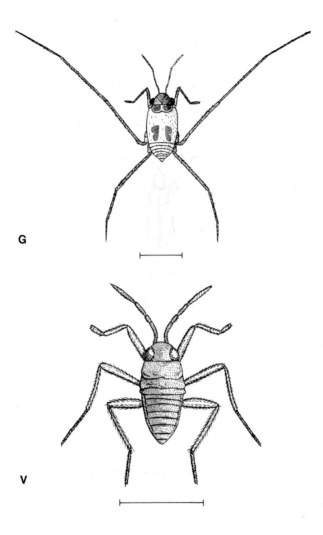

Fig. 6. Nymphs of: *G*, Gerridae (*Gerris najas*); *V*, Veliidae (*Velia* sp.). (Scale lines 1 mm).

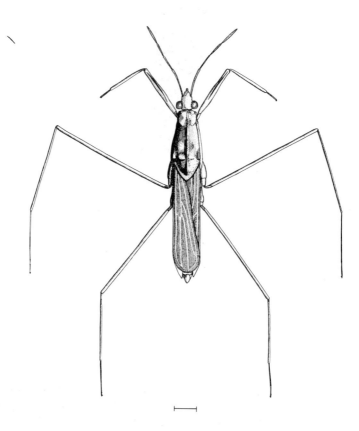

Fig. 7. Gerridae (*Gerris lacustris*). (Scale line 1 mm).

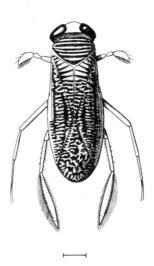

Fig. 8. Corixidae (*Sigara dorsalis*). (Scale line 1 mm).

6(1) Head of a characteristic triangular shape in anterior view with a short, broad rudimentary rostrum (fig. 62); anterior tarsi (palae) always one-segmented and flattened in the male (figs 8, 37) except in two species (fig. 38); scutellum not visible unless the specimen is very small, i.e. 1·7–2·5 mm (fig. 66*pw a*).
Length 1·7–15 mm—   CORIXIDAE, p. 64

— Head with a long, pointed rostrum (fig. 12N); anterior tarsi seldom one-segmented, never flattened; scutellum not concealed by the hemi-elytra (figs 9–12P)—   7

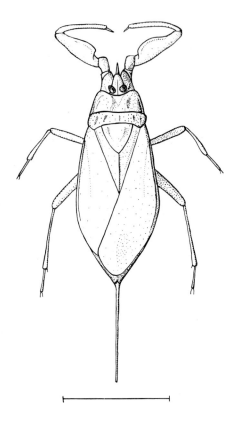

Fig. 9. Nepidae (*Nepa cinerea*). (Scale line 10 mm).

7 Abdomen with two stiff, elongate structures at the posterior end
   forming a single respiratory tube (tail) (figs 9, 10).
   Length 20·0–35·0 mm—   NEPIDAE, p. 53

— No such tubes—   8

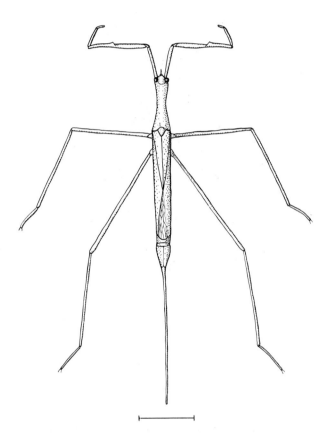

Fig. 10. Nepidae (*Ranatra linearis*). (Scale line 10 mm).

8   Flat and oval animals swimming dorsal surface uppermost; anterior
    legs inserted on the anterior margin of the prosternum (fig. 11)— **9**

—   Boat shaped animals swimming ventral surface uppermost; anterior
    legs inserted on posterior margin of the prosternum (fig. 12)— **10**

9   Head broader than long; rostrum, when folded under the body,
    reaching the bases of the anterior legs; anterior femur very broad with
    tibia folded against it (fig. 11I).
    Length 10·0–15·0 mm—   NAUCORIDAE, p. 53

—   Head as broad as long; rostrum, when folded under the body, reaching
    the bases of the posterior legs; anterior legs not as above (fig. 11A).
    Length 8·0–10·0 mm—   APHELOCHEIRIDAE, p. 53

It is usually possible to separate Naucoridae from Aphelocheiridae on the
presence or absence of wings. So far as I am aware, all recorded specimens
of British Aphelocheiridae were wingless until numerous winged individuals were found in a population in the River Severn (Murray-Bligh 1988).

10(8) Pronotum almost rectangular in dorsal view and, together with the
    hemielytra, covered in coarse pits (fig. 12P); tarsi three-segmented.
    Length 2·0–3·0 mm—   PLEIDAE, p. 64

—   Pronotum wider anteriorly than posteriorly in dorsal view and, toge-
    ther with the hemielytra, smooth (figs 12N, 28); tarsi two-segmented.
    Length 13·0–16·0 mm—   NOTONECTIDAE, p. 54

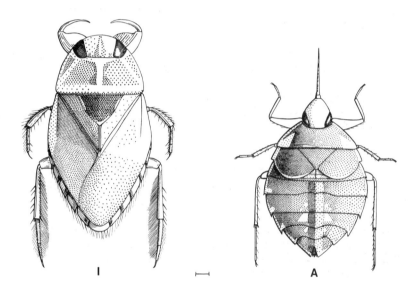

Fig. 11. *I*, Naucoridae (*Ilyocoris cimicoides*); *A*, Aphelocheiridae (*Aphelocheirus aestivalis*). (Scale line 1 mm).

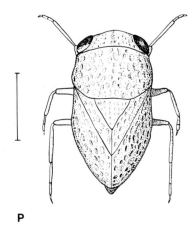

P

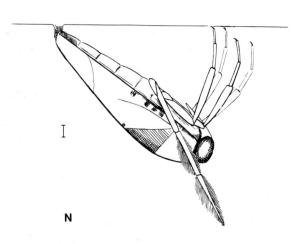

N

Fig. 12. P, Pleidae (*Plea leachi*); N, Notonectidae (*Notonecta glauca*). (Scale lines 1 mm).

KEYS TO SPECIES

Family MESOVELIIDAE

A single species, usually wingless; upper surface with a greenish tinge; the sutures and scattered markings, which may cover most of the dorsal surface, are black (fig. 3)— **Mesovelia furcata** Mulsant & Rey

Family HEBRIDAE

1 First (proximal) segment of antenna at least twice as long as the second; head and pronotum black or brown; membrane of hemielytra smoky with a white spot on each side and a white median streak near the posterior margin; winged.
Length 1·5–2·0 mm— **Hebrus pusillus** (Fallén)

— First segment of antenna only slightly longer than the second; head and anterior part of pronotum reddish brown; membrane of hemielytra without spots; wings usually rudimentary (fig. 4H).
Length 1·2–1·5 mm— **Hebrus ruficeps** (Thomson)

Occasionally, specimens of Hebridae may be collected with semi-terrestrial bugs of the family Dipsocoridae (Leptopodidae), which are of similar size and also occur amongst *Sphagnum* spp.. Dipsocoridae possess only four antennal segments, the proximal two being short and thick whereas the distal two are long and filamentous.

Family HYDROMETRIDAE

All specimens usually have rudimentary hemielytra and no wings.

1   Sixth (sometimes seventh also) visible abdominal segment with a pair of teeth, one placed ventro-laterally on each side; seventh visible abdominal segment emarginate ventrally, so that the posterior margin slopes backwards ventro-dorsally in lateral view (fig. 13*st m, gr m*)—   males 2

—   No abdominal teeth; seventh visible abdominal segment produced posteriorly, ventrally, so that the posterior margin slopes forwards ventro-dorsally in lateral view (fig. 13*st f*)—   females 3

2   Length 9·0–12·0 mm; general colour blackish brown; distance from anterior margin of eyes to anterior tip of head double the distance from posterior margin of eyes to posterior end of head; posterior femora reaching tip of abdomen; sixth and seventh visible abdominal segments each with a pair of teeth; dorsal surface of seventh segment convex so that it terminates in a downwardly directed point in lateral view (fig. 13*st m*)—   **Hydrometra stagnorum** (Linnaeus)

—   Length 7·0–9·0 mm; general colour reddish brown; distance from anterior margin of eyes to anterior tip of head less than double the distance from posterior margin of eyes to posterior end of head; posterior femora reaching the middle of sixth visible abdominal segment; sixth segment only with a pair of teeth; dorsal surface of seventh segment flat and terminating in an upwardly directed point in lateral view (fig. 13*gr m*)—   **Hydrometra gracilenta** Horvath

3(1) Length 9·0–12·0 mm; general colour blackish brown; distance from anterior margin of eyes to anterior tip of head double the distance from posterior margin of eyes to posterior end of head; posterior femora reaching the middle of sixth visible abdominal segment—   **Hydrometra stagnorum** (Linnaeus)

—   Length 7·0–9·0 mm; general colour yellowish brown; distance from anterior margin of eyes to anterior tip of head less than double the distance from posterior margin of eyes to posterior end of head; posterior femora reaching the middle of fifth visible abdominal segment—   **Hydrometra gracilenta** Horvath

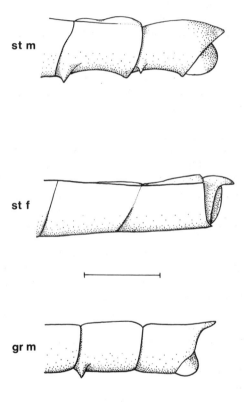

Fig. 13. Hydrometridae: Lateral view of the posterior abdominal segments of *st m*, *Hydrometra stagnorum*, male; *st f*, *H. stagnorum*, female; *gr m*, *H. gracilenta*, male. (Scale line 1 mm).

Family VELIIDAE

All species in this family occur in both winged and wingless forms.
Both forms of *Microvelia* may be identified with confidence. In the genus
Velia both forms of males but only wingless females may be reliably
separated (Brown 1951a). Thus, the key given for winged females is
tentative. Males of *Velia* are best identified by the appearance of the
sclerites of the genital capsule (fig. 15). The last three abdominal segments
should be removed and placed in 50% ethanol on a microscope slide. The
sausage-shaped genital capsule is then dissected out, a drop of glycerol
added which is mixed with the ethanol, and the whole allowed to stand
for a few minutes prior to examination. Alternatively, the genital capsule
may be dehydrated in 100% ethanol, cleared in xylene, and transferred to
cedar wood oil for examination and storage in a microphial.

1   Length 6·0–8·0 mm—   genus VELIA **3**

—   Length 1·2–2·5 mm—   **2**

2   Uniformly black; wingless with metanotum clearly visible (fig. 6V)—
                                                            Immature VELIA

—   Winged (fig. 18), or wingless with only the posterior corners of the
    metanotum visible, the rest being covered by the pronotum and
    mesonotum (fig. 19); patches of white hairs on the thorax and
    abdomen—   genus MICROVELIA **8**

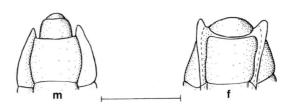

Fig. 14. Veliidae: Dorsal view of the posterior abdominal segments of *Velia saulii*, *m*, male;
f, female. (Scale line 1 mm).

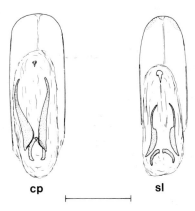

cp   sl

Fig. 15. Veliidae: Dorsal view of the male genital capsule to show sclerites of *cp*, *Velia caprai*; *sl*, *V. saulii*. (Scale line 0·25 mm).

3(1) Two apparent genital segments visible in dorsal view (fig. 14*m*); posterior margins of hind femora with spines, the outermost and innermost being longer than the others (fig. 5)—   males **4**

—   One apparent genital segment visible in dorsal view (fig. 14*f*); hind femora spineless—   females **5**

4   Outer sclerites of genital capsule long and tapering smoothly (fig. 15*cp*)—   **Velia caprai** Tamanini
(Winged and wingless males)

—   Outer sclerites of genital capsule short and tapering suddenly (fig. 15*sl*)—   **Velia saulii** Tamanini
(Winged and wingless males)

5(3) Winged females—   **6**

— Wingless females—   **7**

6   The two proximal white spots on the hemielytra *usually* shorter and
    not overlapping; central spot *usually* oval and larger than distal spot;
    veinlet present on inner margin (fig. 16*cp*)—
    Velia caprai Tamanini
    (Winged females)

—   The two proximal white spots on the hemielytra *usually* longer and
    overlapping; central spot *usually* round and smaller than distal spot;
    veinlet absent but there may be a darker spot in the same place (fig.
    16*sl*)—   **Velia saulii** Tamanini
    (Winged females)

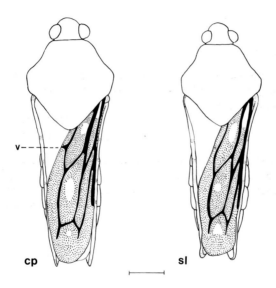

Fig. 16. Veliidae: Dorsal view of winged female to show venation and pattern of the hemielytron of *cp, Velia caprai; sl, V. saulii. (v,* veinlet). Redrawn from Brown (1951a). (Scale line 1 mm).

7(5) First abdominal tergum round and projecting well above the level of the other terga so that, in lateral view, it is seen above the connexiva (see p. 7); connexiva folded inwards so that their edges are concealed in dorsal view; in lateral view the connexiva are straight or sinuous with a straight or upturned posterior end; pronotum not concealing mesonotum (fig. 17*cp*)—   **Velia caprai** Tamanini
(Wingless females)

— First abdominal tergum flat and scarcely projecting above the other terga so that, in lateral view, it is scarcely seen above the connexiva; connexiva standing more or less upright so that the edges may be seen in dorsal view; in lateral view the connexiva are convex with a downturned posterior end (fig. 17*sl*)—   **Velia saulii** Tamanini
(Wingless females)

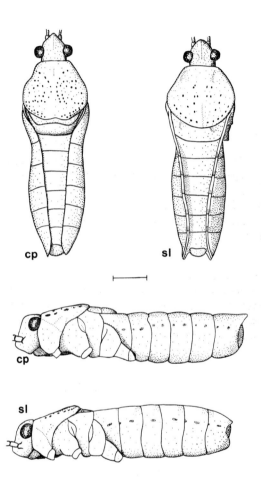

Fig. 17. Veliidae: Dorsal and lateral views of wingless female of *cp*, *Velia caprai*; *sl*, *V. saulii*. (Scale line 1 mm).

8(2) Winged—   **9**

— Wingless—   **11**

9 One (male) or two (female) relatively large oval whitish patches on
   the mid to posterior parts of each hemielytron (fig. 18*bn*); [pronotum
   dark brown; hind tarsus segments unequal, the proximal shorter (fig.
   19*rt t*)].
   Length 1·9–2·1 mm—   **Microvelia buenoi umbricola** Wróblewski
                                             (Winged males and females)

— At least three, usually more, relatively small whitish patches of a
   variety of shapes on the hemielytra (fig. 18*pg*, *rt*)—   **10**

Characters in square brackets are not exclusive.

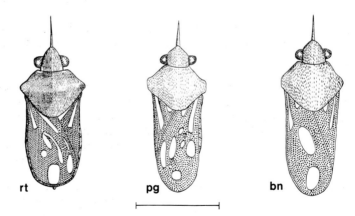

Fig. 18. Veliidae: Dorsal view of winged *rt*, *Microvelia reticulata*; *pg*, *M. pygmaea*; *bn*, *M. buenoi umbricola*. (Scale line 1 mm).

10   Hemielytra dark brown with a central recurved (hooked) whitish patch (fig. 18*rt*); pronotum dark brown, the lateral angles rather pointed; hind tarsus segments unequal (fig. 19*rt t*).
Length 1·6–1·9 mm—   **Microvelia reticulata** (Burmeister)
(Winged males and females)

—   Hemielytra dark grey with a central patch which is not recurved or, if apparently so, is broken into a number of very small patches (fig. 18*pg*); pronotum dark grey, the lateral angles distinctly truncate; hind tarsus segments equal (fig. 19*pg t*).
Length 1.8–2.0 mm—   **Microvelia pygmaea** (Dufour)
(Winged males and females)

11(8) Seven apparently normal abdominal segments in dorsal view without discontinuities in the lateral outline of the animal; in addition, there are one or two distinct genital segments (fig. 19, top row)—
Wingless males **12**

—   Eight apparently normal abdominal segments in dorsal view without discontinuities in the lateral outline; in addition, a much shorter genital segment is visible in one species (fig. 19, middle row)—
Wingless females **14**

Examine the genital segments by looking perpendicularly to the dorsal surface of the posterior end; it is often convexly curved and obscures the genital segments.

12   Pronotum covering mesonotum and metanotum except at sides (fig. 19*pg*); hind tarsus segments equal (fig. 19 *pg t*); [transverse brownish mark near anterior margin of pronotum interrupted in the middle; eighth abdominal segment large and parallel sided].
Length 1·6 mm—   **Microvelia pygmaea** (Dufour)
(Wingless males)

—   Pronotum not covering mesonotum; mesonotum covering metanotum except at sides (fig. 19*bn, rt*); hind tarsus segments unequal, the proximal shorter (fig. 19*rt t*)—   **13**

13   Posterior margins of pronotum and mesonotum in a regular curve and
     parallel; transverse brownish mark near anterior margin of pronotum
     interrupted in middle; eighth abdominal segment small (fig. 19*rt*).
     Length 1·4 mm—   **Microvelia reticulata** (Burmeister)
                                                          (Wingless males)

—   Posterior margins of pronotum and mesonotum sinuate; transverse
     brownish mark near anterior margin of pronotum continuous; eighth
     abdominal segment large but tapering (fig. 19*bn*).
     Length 1·7 mm—   **Microvelia buenoi umbricola** Wróblewski
                                                          (Wingless males)

14(11) Pronotum covering mesonotum and metanotum except at sides (fig.
     19*pg*); hind tarsus segments equal (fig. 19*pg t*; [transverse brownish
     mark near anterior margin of pronotum interrupted in the middle but
     largely obscured by whitish hairs].
     Length 1·8 mm—   **Microvelia pygmaea** (Dufour)
                                                          (Wingless females)

—   Pronotum not covering mesonotum; mesonotum covering metanotum
     except at sides (fig. 19*bn, rt*); hind tarsus segments unequal, the
     proximal shorter (fig. 19*rt t*)—   **15**

15   Posterior margins of pronotum and mesonotum in a regular curve and
     parallel; transverse brownish mark near anterior margin of pronotum
     interrupted in the middle and obscured by whitish hairs (fig. 19*rt*).
     Length 1·6 mm—   **Microvelia reticulata** (Burmeister)
                                                          (Wingless females)

—   Posterior margins of pronotum and mesonotum somewhat sinuate but
     not as clearly as in males; transverse brownish mark near anterior
     margin of pronotum continuous and not obscured by whitish hairs
     (fig. 19*bn*).
     Length 1·7–1·8 mm—   **Microvelia buenoi umbricola** Wróblewski
                                                          (Wingless females)

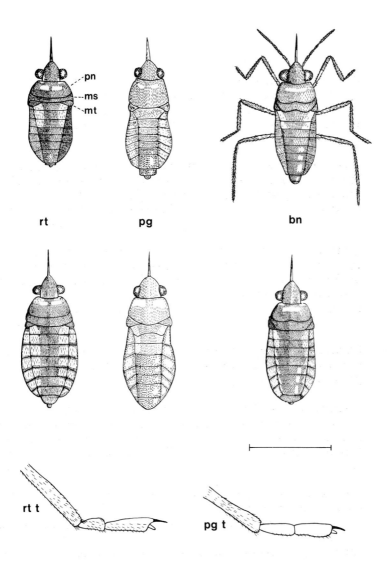

Fig. 19. Veliidae: Dorsal views of wingless males (top row) and females (middle row) of *rt*, *Microvelia reticulata*; *pg*, *M. pygmaea*; *bn*, *M. buenoi umbricola*. (*pn*, pronotum; *ms*, mesonotum, *mt*, metanotum). (Scale line 1 mm). Posterior tarsal segments (bottom row) of *rt t*, *M. reticulata*; *pg t*, *M. pygmaea*. (Scale line 0·25 mm).

Family GERRIDAE

Structural features of taxonomic importance are shown in fig. 20. The pronotum is large, elongate and contains four particular regions: the anterior lobes (a), constriction (c), keel (k) and disk (d). The first abdominal segment is inconspicuous and thus the abdomen appears to contain one segment fewer than the true number. Thus the seventh segment, for example, is designated sixth visible, in parentheses, in the key. The femora of the fore (anterior) legs are ornamented with dark markings which vary in different species (fig. 24lc, od). In the descriptions of the positions of these markings the terms anterior, posterior, upper and lower refer to a leg set in the position shown in figs 7 and 20. The colour of different parts of the insects is valuable in identification but is somewhat variable in fresh specimens and tends to fade in old specimens. It should be noted that teneral (newly emerged) adult specimens of the black species may resemble fully mature specimens of the lighter species. The underside of the body is covered with a pile of fine hairs and hence colour depends on the direction of illumination; this latter point is particularly critical when examining *Gerris lateralis* (see key and fig. 25). In some species there is a broad longitudinal pale area down the ventral surface of the abdomen; this is a pigmentation effect and is not caused by the pattern of hairs.

Males of each species may be distinguished by the characteristic appearance of the sclerites supporting the aedeagus (Poisson 1922). The aedeagus lies in abdominal segment nine (eighth visible) and it may be exposed by removal of the tenth segment and the ninth tergum (figs 20, 21). Alternatively, abdominal segments eight, nine and ten may be removed as a unit and placed in a cavity slide in a drop of 50% ethanol. The egg-shaped aedeagus should be extracted from the ninth segment, dehydrated in 100% ethanol, cleared in xylene, and placed in cedar wood oil on the cavity slide for a few minutes. Then it may be examined and moved to the various positions required (fig. 21lc a–d); fine pins or cabinet points mounted in matchsticks are excellent instruments. After identification it may be stored in cedar wood oil in a microphial attached to the specimen.

The aedeagus contains a maximum of four sclerites which are of complex shapes. It is helpful to familiarise oneself with their three-dimensional structure before attempting comparison with the text-figures since each view is restricted to a convenient focal plane and may not show every aspect of structure. The shapes and spatial relationships are shown in fig. 21lc a–d. The most conspicuous feature in dorsal view is the Y-shaped sclerite (y) which runs down the mid line, the base of the Y being directed posteriorly and enclosed between two further apparent sclerites (x); in

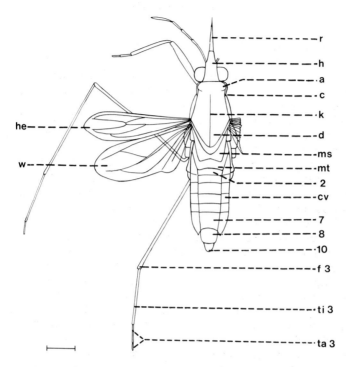

Fig. 20. Gerridae: Dorsal view of *Gerris argentatus*. (*a*, anterior lobe of pronotum; *c*, constriction of pronotum; *cv*, connexivum; *d*, disk of pronotum; *f 3*, femur of posterior leg; *h*, head; *he*, hemielytron; *k*, keel of pronotum; *ms*, mesonotum; *mt*, metanotum; *r*, rostrum (beak); *ta 3*, tarsus of posterior leg; *ti 3*, tibia of posterior leg; *w*, wing; *2*, *7*, *8*, *10*, abdominal segments). (Scale line 1 mm).

fact, they are the forked tip of a single sclerite which curves ventrally and then runs anteriorly (x). There are two additional sclerites (z), one on either side of the central pair (note that sclerite z of the far side has been omitted in fig. 21*lc b–c*). Inspection of figs 21 and 22 shows that, although differing in detail, the sclerites are of the same general arrangement except in *G. lateralis asper*, *G. najas* and *Limnoporus rufoscutellatus*. In *G. lateralis asper*, y is very short, while it is absent in the other two species.

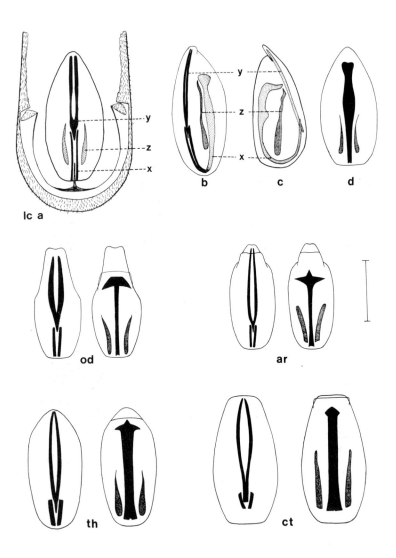

Fig. 21. Gerridae: Dorsal view of ninth sternum and aedeagus, after removal of tenth segment and ninth tergum, of *lc a, Gerris lacustris*. Aedeagus of *G. lacustris* in *b*, dorso-lateral, *c*, lateral, *d*, ventral view. Aedeagus in dorsal (left) and ventral (right) view of *od, G. odontogaster; ar, G. argentatus; th, G. thoracicus; ct, G. costai poissoni*. (*x, y, z*, sclerites of aedeagus – see text). (Scale line 0·25 mm).

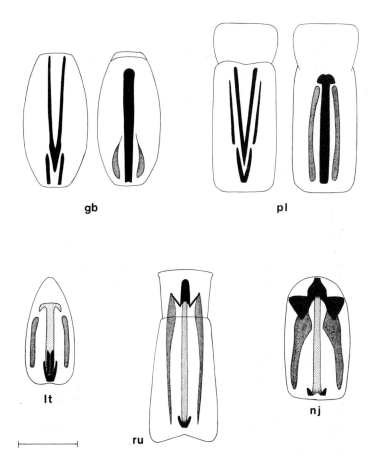

Fig. 22. Gerridae: Aedeagus in dorsal (left) and ventral (right) view of *gb*, *Gerris gibbifer*; *pl*, *G. paludum*. Aedeagus in dorsal view of *lt*, *G. lateralis asper*; *ru*, *Limnoporus rufoscutellatus*; *nj*, *G. najas*. (Scale line 0·25 mm).

Keys to nymphs of Gerridae are given in Brinkhurst (1959c) and Vepsäläinen & Krajewski (1986).

1 Seventh and eighth (sixth and seventh visible) abdominal segments emarginate ventrally and laterally (figs 23*ru*, 24*lc m*, *lt m*, *od m*); seventh segment with a small, distinct additional emargination in some species (fig. 24*lc m*)—   males 2

— Seventh (sixth visible) segment only emarginate, the eighth comprising two valves which meet ventrally to form a keel (figs 23*pl*, *nj*, 24*lc f*, *od f*)—   females 2

From hereon in the key, males and females are considered together.

2 Larger species, length 13·0–18·0 mm; seventh (sixth visible) abdominal segment with long lateral spiniform projections (fig. 23); [seventh segment of male without additional mid-ventral emargination (fig. 23*ru*)]—   3

— Smaller species, length 6·5–14·0 mm; seventh (sixth visible) abdominal segment with only short lateral projections or none (fig. 24); [seventh segment of male with a small mid-ventral emargination except in *G. lateralis asper*, length 9·0–11·0 mm]—   5

Characters in square brackets are not exclusive.

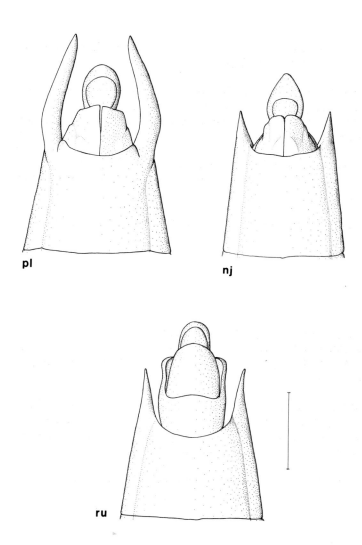

Fig. 23. Gerridae: Ventral view of the posterior abdominal segments of *pl*, *Gerris paludum* female; *nj*, *G. najas* female; *ru*, *Limnoporus rufoscutellatus* male. (Scale line 1 mm).

3   Pronotum reddish, especially on the disk; antennae more than half
    as long as the body; first (proximal) antennal segment shorter than
    second and third together, ratio 3:4; middle femora shorter than the
    hind femora.
    Length 13·0–17·0 mm—   **Limnoporus rufoscutellatus** (Latreille)

—   Pronotum black; antennae less than half as long as the body; first
    (proximal) antennal segment longer than second and third together,
    ratio 4:3; middle femora as long as, or longer than, hind femora—
                                                                       4

4   Sides of pronotum with a yellow line; lateral spiniform projections of
    seventh abdominal segment extending to, or beyond, tip of abdomen
    (fig. 23*pl*); usually winged.
    Length 14·0–16·0 mm—   **Gerris paludum** Fabricius

—   Sides of pronotum uniformly dark; lateral spiniform projections of
    seventh abdominal segment not reaching tip of abdomen (fig. 23*nj*);
    usually wingless.
    Length 13·0–18·0 mm—   **Gerris najas** Degeer

5(2) Pronotum yellowish or reddish on disk; at least proximal segments
    of antennae dark yellowish or reddish brown—   **6**

—   Pronotum black on disk, sometimes with a narrow pale longitudinal
    stripe between the anterior lobes; antennae black—   **8**

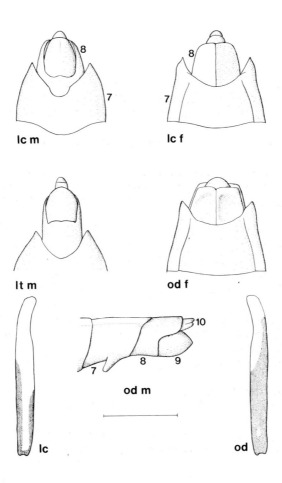

Fig. 24. Gerridae: Ventral view of the posterior abdominal segments of *lc m*, *Gerris lacustris* male; *lc f*, *G. lacustris* female; *lt m*, *G. lateralis asper* male; *od f*, *G. odontogaster* female. Lateral view of posterior abdominal segments of *od m*, *G. odontogaster* male. (7-10, abdominal segments). Anterior view (see text) of the fore femur of *lc*, *G. lacustris*; *od*, *G. odontogaster* (Scale line 1 mm).

Fig. 25. Gerridae: Ventral view of the abdomen of *Gerris lateralis asper* to show patches of
silver hairs. (Scale line 1 mm).

6   Underside of abdomen covered with silver-grey hairs, at the sides two
     lines of bright silvery hairs (fig. 25); fore femora dark yellowish brown
     with a single broad black stripe on the upper and posterior surface;
     seventh abdominal segment of male without a small mid-ventral
     emargination (fig. 24*lt m*).
     Length 9·0–11·0 mm—   **Gerris lateralis asper** (Fieber)

—   Underside of abdomen uniformly covered with silver-grey hairs; fore
     femora light with two black stripes, one as above and a second narrow
     stripe on the under surface extending from the tip midway to the base
     (proximal end); seventh abdominal segment of male with a small mid-
     ventral emargination (fig. 24*lc m*)—   7

The abdomen should be pointed directly towards the light source for
examination of the abdominal hairs. There may be some difficulty at first
but the lines of hairs are exclusive to *Gerris lateralis* among Gerridae.

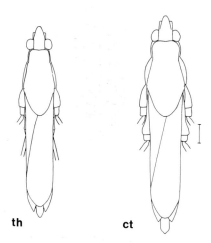

Fig. 26. Gerridae: Dorsal view of body outline of *th, Gerris thoracicus*; *ct, G. costai poissoni.* (Scale line 1 mm).

7 Length 10·0–12·0 mm; body more or less parallel-sided (fig. 26*th*); disk of pronotum, narrow line between anterior lobes and line on side of pronotum behind constriction yellowish brown—

Gerris thoracicus Schummel

— Length 12·0–14·0 mm; body broad in the region of the middle coxae (fig. 26*ct*); disk of pronotum, narrow line between anterior lobes, line on side of pronotum behind constriction and small, often obscure, mark on side of pronotum before constriction reddish brown—

Gerris costai poissoni Wagner & Zimmermann

Fig. 27. Gerridae: Lateral view of the pronotum of *lc*, *Gerris lacustris*; *gb*, *G. gibbifer*. (Scale line 1 mm).

8(5) Length 10·0–13·0 mm; metasternum with a yellowish tubercle; yellow line on side of pronotum not continued in front of constriction (fig. 27*gb*); [centre of underside of abdomen pale; markings on fore femora as in fig. 24*od*]—   **Gerris gibbifer** Schummel

—   Length 6·5-10·0 mm; metasternum without a tubercle; yellow line on side of pronotum continued in front of constriction (fig. 27*lc*)—   **9**

9   Fore femora pale with two narrow black stripes stretching back from their distal ends, one on the upper-anterior surface and a shorter one on the under-anterior surface (fig. 24*lc*); [female with a pale band down the centre of the underside of the abdomen and a strongly marked keel on the eighth segment (fig. 24*lc f*).
Length 8·0-10·0 mm—   **Gerris lacustris** (Linnaeus)

—   Fore femora pale proximally but distal two-thirds almost entirely black (fig. 24*od*)—   **10**

10   Hind tibiae and tarsi together almost as long as the femora; no silvery hairs on the pronotum; male with a ventro-lateral tooth on each side of the seventh abdominal segment (fig. 24*od m*); female with a poorly developed keel on the eighth segment and depressions at its base (fig. 24*od f*).
Length 7·0-9·0 mm—   **Gerris odontogaster** (Zetterstedt)

—   Hind tibiae and tarsi together about two-thirds as long as the femora; a line of silvery hairs about the base of the disk posteriorly; male without abdominal teeth; female with a strongly marked keel on the eighth abdominal segment (as fig. 24*lc f*).
Length 6·5-8·0 mm—   **Gerris argentatus** Schummel

Family NEPIDAE

1 Body long, oval and dorso-ventrally flattened (fig. 9, p. 24).
Length 17·0–23·0 mm— **Nepa cinerea** Linnaeus

— Body long, narrow and cylindrical (fig. 10, 25).
Length 30·0–35·0 mm— **Ranatra linearis** (Linnaeus)

Family NAUCORIDAE

A single species. Pronotum brownish, flecked with dark brown; scutellum black; hemielytra dark olive brown (fig. 11I, 27).
Length 11·0–15·0 mm— **Ilyocoris cimicoides** (Linnaeus)

Family APHELOCHEIRIDAE

A single species. Antennae yellowish brown; hemielytra usually short (fig. 11A, 27).
Length 8·0–10·0 mm— **Aphelocheirus aestivalis** (Fabricius)

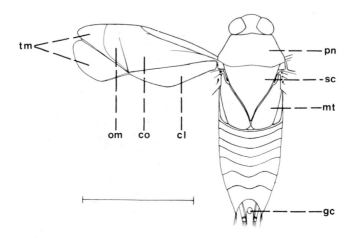

Fig. 28. Notonectidae: Dorsal view of *Notonecta glauca*. (*cl*, clavus; *co*, corium; *gc*, genital
   capsule; *mt*, metanotum; *om*, opaque region of membrane; *pn*, pronotum; *sc*,
   scutellum; *tm*, translucent region of membrane). (Scale line 10 mm).

Family NOTONECTIDAE

Structural features of taxonomic importance are shown in fig. 28. The
key comprises three sections; one based on superficial characters and two
others based on the male and female genitalia respectively. The key based
on female genitalia is somewhat tentative and offered as a means of progress
rather than as a complete solution. Walton (1936) has illustrated further
features and Poisson (1933) placed taxonomic importance on sclerite *k* of
the ovipositor (fig. 31*fg*)
   The majority of specimens may be readily identified by their superficial
characters. However, teneral (newly emerged) specimens, where the pig-
ment is not fully developed, may present problems and fully developed
pigmented specimens may be found with unusual hemielytral markings.
These difficulties may be resolved by examination of the genitalia.
   The male genital capsule should be fully exposed by removal of the
posterior abdominal terga. It should be removed complete, boiled for a
few minutes in 5% potassium hydroxide solution (take care!), washed in
water and placed on a slide in 50% ethanol. The position of the aedeagus
and parameres should be determined (fig. 31*mg*). These structures should

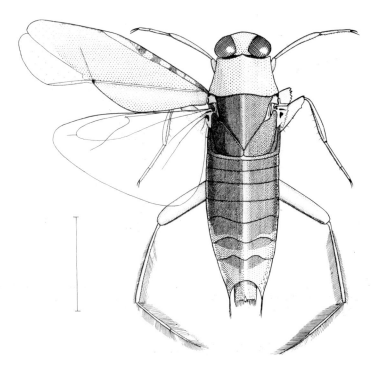

Fig. 29. Notonectidae: Dorsal view of *Notonecta glauca*. (Scale line 5 mm).

be carefully dissected away and the remaining debris brushed aside. The key for male genitalia may be used, starting at couplet 5.

In contrast, the female genitalia are best approached from the ventral surface. The eighth sternum should be removed, taking care to keep it intact (fig. 33); the ovipositor will be seen sufficiently clearly without further dissection (fig. 31*fg*). The key for female genitalia may be used by starting at couplet 8. The long hairs on the sternum must be carefully deflected for comparison with fig. 33.

For both males and females, the genital structures that have been removed may be made into permanent slides.

1 Posterior end of abdomen truncate in lateral view; edges of posterior-most segments sinuate in lateral view; rounded genital capsule often visible (fig. 31*m*)— males **2** or **5**

— Posterior end of abdomen smoothly rounded from dorsal surface to the point where it meets the eighth sternum; edges of posterior-most segments following a smooth curve (fig. 31*f*); ovipositor (fig. 31*fg*) visible if the eighth sternum is deflected aside— females **2** or **8**

2 Length usually less than 14·5 mm; anterior angles of pronotum pointed in both dorsal and lateral views, embracing eyes; dark markings along the anterior margin of the corium and along the base of the membrane (fig. 30*mm*).
Length 13·0–15·0 mm— **Notonecta marmorea viridis** Delcourt

— Length usually more than 14·5 mm; anterior angles of pronotum obtuse, not embracing eyes (fig. 29); dark markings on hemielytra not as above— **3**

3 Metanotum orange, except for two black spots which are seen if the hemielytra are deflected sideways; hemielytra generally mottled with dark markings (fig. 30*mc*).
Length 14·0–16·0 mm— **Notonecta maculata** Fabricius

— Metanotum black (fig. 29); mottled markings on hemielytra not usually present (figs 29, 30*ob*), if so, they are normally confined to the anterior margin of the corium although they may be more extensively distributed in some specimens— **4**

4 Hemielytra pale with irregular dark patches along the anterior edge of the corium (fig. 29); sometimes with additional mottling.
Length 14·0–16·0 mm— **Notonecta glauca** Linnaeus

— Hemielytra dark except for two pale longitudinal bands, one on the corium and one on the clavus (fig. 30*ob*).
Length 14·0–16·0 mm— **Notonecta obliqua** Gallen

— As above, but with an additional pale spot beyond the corial band— **Notonecta obliqua** var. **delcourti** Poisson

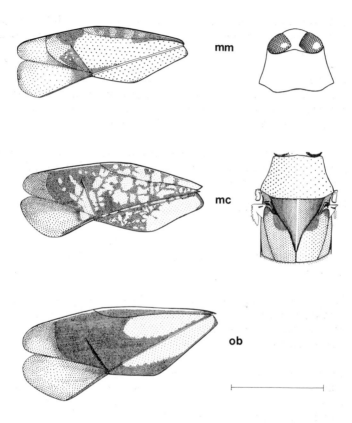

mm

mc

ob

Fig. 30. Notonectidae: Dorsal view of hemielytron and thoracic region of *mm*, *Notonecta marmorea viridis*; *mc*, *N. maculata*. Hemielytron of *ob*, *N. obliqua* (Scale line 5 mm).

MALE GENITALIA:

5(1) Posteriorly-directed dorsal process of the terminal sclerite of the aedeagus long and forming an angle of 70–90° with the body of that structure in lateral view (figs 31*mg*, 32*a gl, a ob*); ventral processes of paramere forming a complete semicircle (fig. 32*p gl, p ob*)— **6**

— Dorsal process of the terminal sclerite of the aedeagus short and upwardly directed, or absent, in lateral view (figs 31*mg*, 32*a mm*); ventral processes of paramere forming an open crescent (fig. 32*p mm, p mc*)— **7**

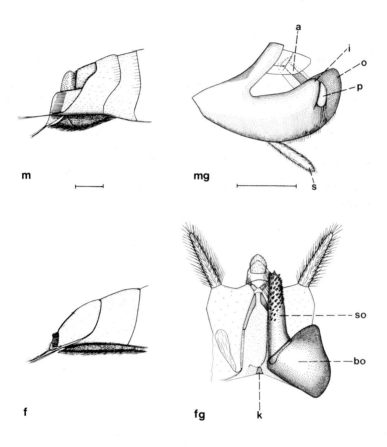

Fig. 31. Notonectidae: Lateral view of the posterior abdominal segments of *m, Notonecta glauca* male; *f, N. glauca* female. Lateral view of the genital capsule of *mg, N. glauca* male. (*a*, terminal sclerite of aedeagus; *i*, inner lobe of capsule; *o*, outer lobe; *p*, paramere; *s*, style). Ventral view of genitalia, with eighth sternum and one ovipositor removed, of *fg, N. glauca* female. (*bo*, base of ovipositor; *k*, sclerite *k*; *so*, shaft of ovipositor). (Scale lines 1 mm).

6   Dorsal process of terminal sclerite of aedeagus slightly curved ventr-
    ally in lateral view (fig. 32*a gl*; posterior end of paramere truncate
    (fig. 32*p gl*)—   **Notonecta glauca** Linnaeus

—   Dorsal process of terminal sclerite of aedeagus virtually straight
    in lateral view (fig. 32*a ob*); posterior end of paramere continuing
    smoothly to a rounded tip (fig. 32*p ob*)—
                                                  Notonecta obliqua Gallen

7(5) Dorsal part of terminal sclerite of aedeagus with lateral processes so
    that it appears **T**-shaped in posterior view; ventral surface of posterior
    part of paramere markedly sinuate and ending in a point (fig.
    32*p mc*)—   **Notonecta maculata** Fabricius

—   Terminal sclerite of aedeagus short and without lateral processes (fig.
    32*a mm*); ventral surface of posterior part of paramere only slightly
    curved and ending in a rounded tip (fig. 32*p mm*)—
                                        Notonecta marmorea viridis Delcourt

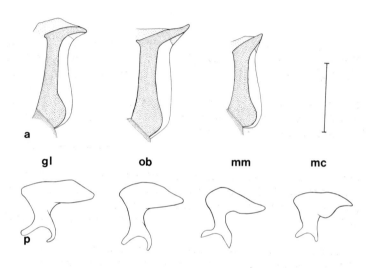

Fig. 32. Notonectidae: Lateral outline views of *a*, the terminal sclerite of the aedeagus and *p*, a paramere of male of *gl, Notonecta glauca*; *ob. N. obliqua*; *mm, N. marmorea viridis*; *mc, N. maculata* (paramere only). (Scale line 0·5 mm).

FEMALE GENITALIA:

8(1) Eighth sternum with an emarginate tip (fig. 33*mc*); ovipositor as in
      fig. 33*mc*—   **Notonecta maculata** Fabricius

—   Eighth sternum with a pointed tip (fig. 33*mm, gl, ob*); ovipositor as
      in fig. 31*fg*—   **9**

 9   Pointed tip of eighth sternum rather straight sided; postero-lateral
      regions slightly recurved towards the centre (fig. 33*ob*)—
                                              Notonecta obliqua Gallen

—   Pointed tip of eighth sternum with convexly curved sides; postero-
      lateral regions not recurved (fig. 33*mm, gl*)—   **10**

10   Pointed tip extending more than one-fifth of the total length of the
      eighth sternum (fig. 33*gl*)—   **Notonecta glauca** Linnaeus

—   Pointed tip extending less than one-fifth of the total length of the
      eighth sternum (fig. 33*mm*)—
                                   Notonecta marmorea viridis Delcourt

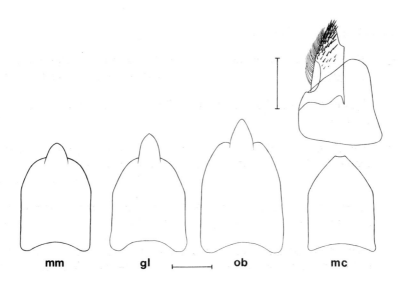

Fig. 33. Notonectidae: Ventral view of eighth sternum of female of *mm, Notonecta marmorea viridis; gl, N. glauca; ob, N. obliqua.* Eighth sternum and ovipositor of *mc, N. maculata.* (Upper scale line 0·5 mm; lower 1 mm).

Family PLEIDAE

A single species. Length 1·8–2·8 mm; yellowish white to whitish grey, underside brownish (fig. 12*P*, p. 28)—

Plea leachi McGregor & Kirkaldy

Family CORIXIDAE

The key applies to both sexes but, in almost every species in this family, males may be more easily identified than females. Accordingly, specimens should be collected in reasonable numbers and sexed (see couplet 2) before being examined further. Males should be identified first and then females may be matched with known males. Care is needed, for the sexes are of different sizes in some species, for example *Sigara lateralis*, but the hemielytral patterns are of considerable assistance (figs 42, 48, 49, 56, 60, 65).

With some experience, solitary females may be identified directly from the key but separation is particularly difficult between five pairs of species: *Corixa punctata* and *C. iberica*, *C. affinis* and *C. panzeri*, *Sigara stagnalis* and *S. selecta*, *Hesperocorixa castanea* and *H. moesta*, *S. dorsalis* and *S. striata* — separation of the last pair is well nigh impossible! Höregott & Jordan (1954) provide a key for the identification of females but I have encountered difficulties with British specimens.

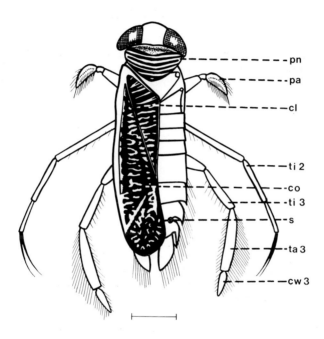

Fig. 34. Corixidae: Dorsal view of *Sigara scotti* male. (*cl*, clavus; *co*, corium; *cw 3*, claw of posterior leg; *pa*, pala (tarsus of anterior leg); *pn*, pronotum; *s*, strigil; *ta 3*, tarsus of posterior leg; *ti 2*, tibia of middle leg; *ti 3*, tibia of posterior leg). (Scale line 1 mm).

The basic features of taxonomic importance are shown in figs 34–37. The anterior tarsi (palae) of males of the subfamily Corixinae are characteristic in almost every species, their shape and the arrangement of pegs providing valuable diagnostic characters. These are not always described in the key, since words are somewhat inadequate, but reference should be made to the text-figures (figs 44, 47, 50, 52, 58, 61, 64). If there is still uncertainty, the parameres should be examined; on rare occasions this latter procedure is essential. The male genital capsule lies inside the posterior end of the abdomen (fig. 35). It may be dissected out by removal of the sterna or, if already protruding, levered out with a needle; it should be placed on a microscope slide in 50% ethanol as in fig. 35*fl*. The right paramere, usually the palest of the various structures present, should be removed and compared with the text-figures (figs 41, 44, 47, 50, 52, 53, 57, 58, 61, 64). Note that in the genus *Corixa* the asymmetry is reversed and both parameres should be removed (figs 35, 41). Both parameres should also be removed in Micronectinae (fig. 66). Permanent slides may be made from these dissections.

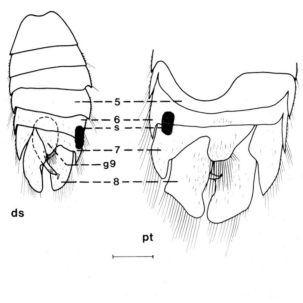

ds

pt

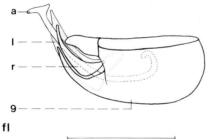

fl

Fig. 35. Corixidae: Dorsal view of the abdomen of male *ds*, *Sigara dorsalis*; *pt*, *Corixa punctata*. (*g 9*, genital capsule and segment 9; *s*, strigil; *5–8*, abdominal segments). Right lateral view, after removal, of genital capsule of male *fl*, *Sigara falleni*. (*a*, aedeagus; *l*, left paramere; *r*, right paramere; *9*, segment 9). (Scale lines 1 mm).

The direction of illumination in relation to the specimen is particularly critical at some couplets in the key, and the relevant comments *must* be noted.

1   Segmentation of the abdomen clearly visible in dorsal view; wing buds present in larger specimens—   nymphs

For identification, see Cobben & Pillot (1960) and Jansson (1969).

—   Dorsal surface of the abdomen covered by the hemielytra (figs 8, 34)—   adults 2

2   Anterior and posterior margins of the abdominal segments irregularly curved, never parallel (figs 35*ds*, *pt*, 36*m*); front of head with a flat or concave area (e.g. fig. 62, p. 103); in Corixinae the pala is flat and has one or two rows of strong, short pegs (figs 37, 44, 47, 50, 52, 58, 61, 64)—   males 3

—   Anterior and posterior margins of the abdominal segments straight, or regularly curved, and parallel (fig. 36*f*); front of head convex unless there are long facial hairs; in Corixinae the pala is triangular in cross section and devoid of pegs (fig. 37*f*)—   females 3

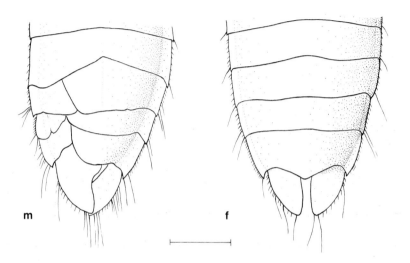

Fig. 36. Corixidae: Ventral view of the abdomen of *Sigara falleni; m*, male; *f*, female. (Scale line 1 mm).

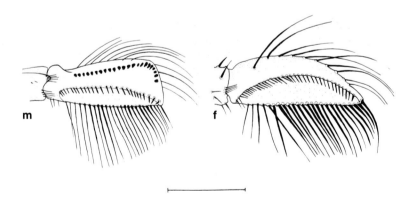

Fig. 37. Corixidae: Anterior view of the pala (anterior tarsus) of *Hesperocorixa sahlbergi; m*, male; *f*, female. (Scale line 0·5 mm).

From hereon in the key, males and females are considered together.

3　Length 3·0–14·0 mm; scutellum covered by pronotum (fig. 34); antennae with four segments—　　　　　　　　　　　　　　　　　**4**

—　Length 1·7–2·5 mm; scutellum visible (fig. 66*pw a*, p. 111); antennae with three segments—　　　　　subfamily MICRONECTINAE **35**

4　Pronotum uniformly brown, occasionally with a slight trace of transverse dark lines; anterior tarsus long and cylindrical (fig. 38). Length 3·0–6·5 mm—　　　　　　subfamily CYMATIAINAE **5**

—　Pronotum with alternate light and dark transverse lines (fig. 34); anterior tarsus (pala) short, flat in male and triangular in female (fig. 37). Length 4·5–14·0 mm—　　　　　　subfamily CORIXINAE **6**

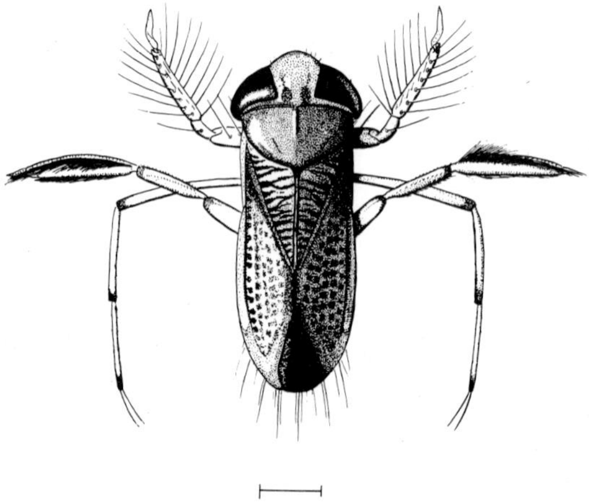

Fig. 38. Corixidae: Dorsal view of *Cymatia bonsdorffii*. (Scale line 1 mm; drawing by G. A. Walton).

5 Length 5·5–6·5 mm; clavus with light and dark transverse lines;
   corium with an irregular light and dark pattern; pronotum about
   twice as broad as long (fig. 38)—
                              Cymatia bonsdorffii (Sahlberg, C.)

— Length 3·0–4·5 mm; clavus usually a uniform dark brown; corium
   brown with two darker longitudinal lines; pronotum about three
   times as broad as long—   **Cymatia coleoptrata** (Fabricius)

6(4) Larger species, least breadth 3·5 mm, length 8·0-14·0 mm; pronotum
      and hemielytra smooth and shiny; strigil of male on the left side of
      the abdomen; cleft on the dorsal surface of the seventh segment on
      the right side (fig. 35*pt*)—   7

— Smaller species, not more than 3·0 mm broad, length 4·5–10·0 mm;
   pronotum and, except in three small species, clavus and corium of
   the hemielytra finely wrinkled; strigil, if present, on the right side of
   the abdomen; cleft on the dorsal surface of the seventh segment on
   the left side (fig. 35*ds*)—   11

The distinction between smooth, shiny surfaces and wrinkled surfaces
often presents difficulty to beginners. The surface should be dry. The
longitudinal axis of the specimen should be placed at right angles to a
good light source and used as a mirror to reflect light up a low-power
microscope with a magnification of ×15. A so-called smooth, shiny surface
may have some hairs or pits but never has the characteristic longitudinal
rippled appearance of the wrinkled surface, like a sandy shore after the
tide has receded.

7   Length usually 12·0–14·0 mm; middle claws shorter than tarsi; usu-
    ally (14) 15–20 pronotal lines—   **8**

—   Length usually 8·0–11.0 mm; middle claws as long as or longer than
    tarsi; usually 10-14 pronotal lines—   **10**

The claws and tarsi should be measured as their differences in width tend
to cause errors if judged by eye alone.

8   Proximal (inner) end of middle tibia tapering slightly at the point
    where it articulates with the femur (fig. 39*pt*); body of right paramere
    about one-third as deep as long and with a distinct narrow upwardly-
    pointing terminal process (fig. 41*ib, pt*)—   **9**

—   Proximal (inner) end of middle tibia becoming suddenly narrower
    just before the point where it articulates with the femur so that there
    is a distinct bay in the outline of the lower surface, in the male there
    is a tuft of bristles on the end of the femur (fig. 39*dn*); body of right
    paramere of male about one-quarter as deep as long and without an
    upwardly-pointing terminal process (fig. 41*dn*).
    Length 12.0-13.0mm—   **Corixa dentipes** (Thomson)

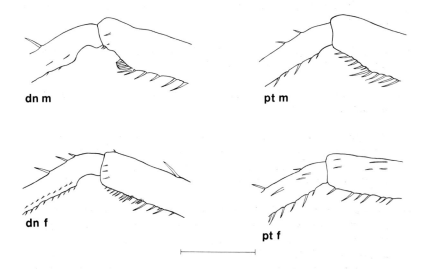

Fig. 39. Corixidae: Anterior view of the junction of the femur and tibia of the middle leg of
dn m, *Corixa dentipes* male; *dn f*, *C. dentipes* female; *pt m*, *C. punctata* male; *pt m*,
C. punctata male; *pt f*, *C. punctata* female. (Scale line 1 mm).

9 Middle femur of male slightly curved when seen in postero-ventral
   view and with a row of spines forming a curve (fig. 40*pt m*); right
   paramere with two distinct dorsal processes, one terminal and one
   central (fig. 41*pt*).
   Length 12·0–14·0 mm—   **Corixa punctata** (Illiger)

— Middle femur of male straight when seen in postero-ventral view and
   with a straight row of spines (fig. 40*ib m*); right paramere with a
   distinct terminal dorsal process but the central process is much
   reduced (fig. 41*ib*).
   Length 12·0–14·0 mm—   **Corixa iberica** Jansson

Jansson (1981, 1986) separates the females of these species on the basis of
a shallow curved ridge on the middle femur of *C. punctata* and its absence
in *C. iberica*. The ridge is poorly developed in some British females of *C.
punctata* and this feature should be used with caution (fig. 40*pt f, ib f*).

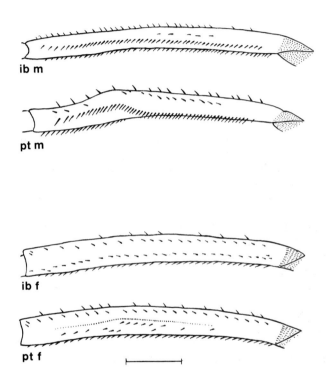

Fig. 40. Corixidae: Postero-ventral view of the femur of the middle leg of *ib m*, *Corixa iberica* male; *pt m*, *C. punctata* male. Antero-ventral view of the femur of the middle leg of *ib f*, *C. iberica* female; *pt f*, *C. punctata* female. (Scale line 1 mm).

10(7) Length usually 10·0–11·0 mm; dark lines on hemielytra generally broader than light ones; male pala with more than 30 pegs; body of right paramere about one-quarter as deep as long (fig. 41*pz*)—
Corixa panzeri (Fieber)

— Length usually 8·0-9·0 mm; dark lines on hemielytra generally narrower than light ones; male pala with fewer than 30 pegs; body of right paramere very narrow, about one-eighth as deep as long (fig. 41*af*)—   **Corixa affinis** Leach

11(6) Posterior tarsi with a dark mark at the distal end or posterior claws dark (fig. 63)—   **32**

— Posterior tarsi uniformly pale (fig. 34)—   **12**

The long hairs along the margin of the tarsi should be brushed aside as they can have the appearance of a dark mark.

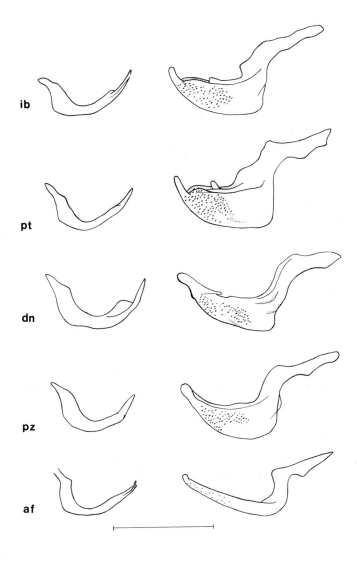

ib

pt

dn

pz

af

Fig. 41. Corixidae: Left and right parameres of male *ib, Corixa iberica; pt, C. punctata; dn, C. dentipes; pz, C. panzeri; af, C. affinis.* (Scale line 1 mm).

12 Light lines on corium very narrow and mostly extending unbroken
    across the black background (fig. 42*ln, sb*), or the entire dorsal surface
    brownish or chestnut coloured with little contrast between light and
    dark (fig. 42*cs, ms*); metasternal xiphus distinctly longer than its
    breadth at the base (fig. 43*sb*); male palae of a consistent characteristic
    shape (fig. 44*sb p, cs p*)—   **13**

—   Light lines on corium generally broader and more broken (e.g. figs.
    49, 60); specimens never brownish with little contrast between light
    and dark; metasternal xiphus either no longer or shorter than its
    breadth at the base (fig. 43*ds, lm, gr, dt*) except in *Sigara semistriata*
    (fig. 43*sm*); male palae not as above (figs. 47, 50, 57, 58, 61)— **16**

This couplet may present problems. Some specimens of *Sigara falleni* and
S. distincta have relatively narrow transverse lines on the hemielytra (fig.
49) and may be incorrectly determined if this criterion is used alone.
However, there should be no problem if used in combination with the
shape of the metasternal xiphus.

13 Length 7·0–9·0 mm; pronotum and hemielytra black or nearly so with
    sharply contrasted yellow lines running across them (fig. 42*ln, sb*)—
                                                                        14

—   Length 4·5–6·0 mm; pronotum and hemielytra brownish or chestnut
    coloured with little contrast between light and dark areas (fig.
    42*cm, ms*)—   **15**

14 Pronotum with 6 pale transverse lines and a fine light line round the
    margin; dark markings on hemielytra extending to posterior apex of
    corium (fig. 42*ln*).
    Length 7·0–8·0 mm—   **Hesperocorixa linnaei** (Fieber)

—   Pronotum with 7–9 pale transverse lines and a uniformly dark margin;
    dark markings on hemielytra not quite extending to posterior apex of
    corium (fig. 42*sb*).
    Length 7·0–9·0 mm—   **Hesperocorixa sahlbergi** (Fieber)

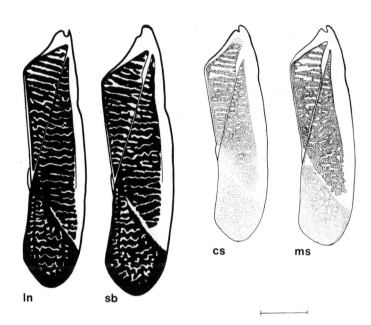

Fig. 42. Corixidae: Hemielytron of *ln*, *Hesperocorixa linnaei*; *sb*, *H. sahlbergi*; *cs*, *H. castanea*; *ms*, *H. moesta*. (Scale line 1 mm).

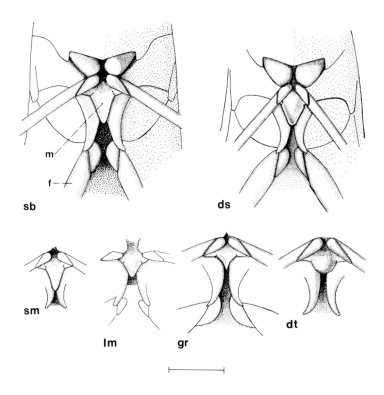

Fig. 43. Corixidae: Ventral view of the abdomen, to show the metasternal xiphus, of *sb*, *Hesperocorixa sahlbergi*; *ds*, *Sigara dorsalis*; *sm*, *S. semistriata*; *lm*, *S. limitata*; *gr*, *Arctocorisa germari*; *dt*, *S. distincta*. (*f*, femur of posterior leg; *m*, metasternal xiphus). (Scale line 1 mm).

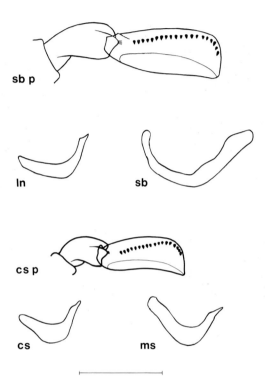

Fig. 44. Corixidae: Pala of male *sb p*, *Hesperocorixa sahlbergi*; *cs p*, *H. castanea*. Right paramere of male *ln*, *H. linnaei*; *sb*, *H. sahlbergi*; *cs*, *H. castanea*; *ms*, *H. moesta*. (Scale line 0·5 mm).

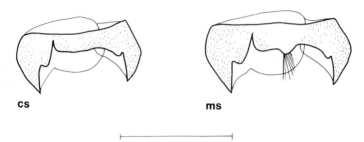

Fig. 45. Corixidae: Dorsal view of the seventh abdominal segment of male *cs*, *Hesperocorixa castanea*; *ms*, *H. moesta*. (Scale line 1 mm).

15(13) Middle claws same length or shorter than tarsi; posterior dorsal margin of seventh abdominal segment of male with a tuft of hairs (fig. 45*ms*).
Length 5·5–6·0 mm—   **Hesperocorixa moesta** (Fieber)

— Middle claws longer than tarsi; posterior dorsal margin of seventh abdominal segment of male without a tuft of hairs (fig. 45*cs*).
Length 4·5–5·5 mm—   **Hesperocorixa castanea** (Thomson)

The male palae are of little diagnostic value within the genus *Hesperocorixa*. However, the four species may be separated by removal and inspection of the right paramere (fig. 44).

16(12) Corium smooth and shiny; concavity on front of male head is interrupted by a transverse ridge (fig. 46); [Length 5·0–6·5 mm]—
                                                                                    17

— Corium wrinkled; if there is a transverse ridge on front of male head, the concavity is only present ventral to the ridge (fig. 62*ng*)—   **18**

See note to couplet 6 (p. 71).

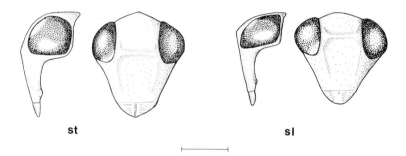

st   sl

Fig. 46. Corixidae: Lateral and anterior views of the head of male *st, Sigara stagnalis; sl, S. selecta* (Scale line 1 mm).

17 Transverse ridge on front of male head situated about midway between the upper and lower edges of the eyes (fig. 46*sl*); pronotal keel extending from anterior margin to 4th or 5th transverse black line. Length 5·2–6·2 mm—   **Sigara selecta** (Fieber)

—   Transverse ridge on front of male head situated just above the level of the lower edge of the eyes (fig. 46*st*); pronotal keel extending to 3rd (occasionally 4th) transverse black line. Length 6·0–6·5 mm—   **Sigara stagnalis** (Leach)

Males of these two species are best separated on the position of the facial transverse ridge and the palae (figs 46, 47). Females can only be separated on the length of the pronotal keel and some specimens cannot be satisfactorily separated at all. The longitudinal pronotal keel can be seen only when the specimen is illuminated directly from the side so that one side of the keel is fully illuminated and the other is in deep shadow. The hemielytra are of value in matching females with known males (fig. 48).

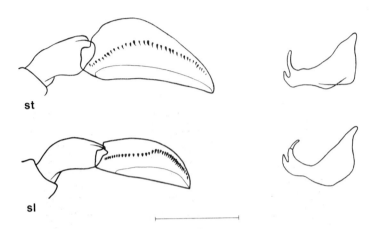

Fig. 47. Corixidae: Pala and right paramere of male *st*, *Sigara stagnalis*; *sl*, *S. selecta* (Scale line 0·5 mm).

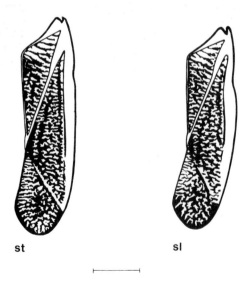

Fig. 48. Corixidae: Hemielytron of *st*, *Sigara stagnalis*; *sl*, *S. selecta* (Scale line 1 mm).

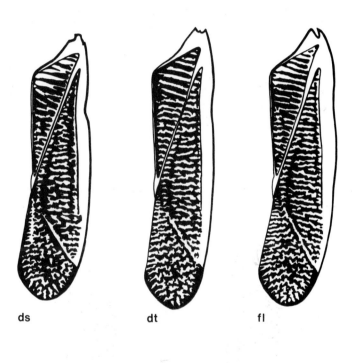

ds   dt   fl

Fig. 49. Corixidae: Hemielytron of *ds*, *Sigara dorsalis*; *dt*, *S. distincta*; *fl*, *S. falleni* (Scale line 1 mm).

18(16) Length 6·5–10·0 mm—   **19**

— Length 4·5–6·75 mm—   **27**

This size distinction works extremely well provided specimens are measured accurately. Specimens which fall into the narrow zone of overlap
should be followed along both parts of the couplet; any error will soon
be revealed.

19 Length 6·5–9·0 mm; markings on corium in rather regular transverse
   series (fig. 49); pronotal keel short, rarely one-quarter of the length
   of the pronotum; metasternal xiphus terminating in a round point
   (fig. 43*ds*, *dt*); male pala not more than twice as long as breadth at
   mid point, two rows of pegs although one of these is poorly developed
   in one species (fig. 50)—   **20**

— Length 7·5–10·0 mm; markings on corium irregular (fig. 56); pronotal
   keel long, virtually the whole length of the pronotum; metasternal
   xiphus terminating in a sharp point (fig. 43*gr*); male pala about three
   times as long as breadth at mid point, one row of pegs (figs 57, 58)—
                                                                      24

See note to couplet 17 for illumination of pronotal keel.

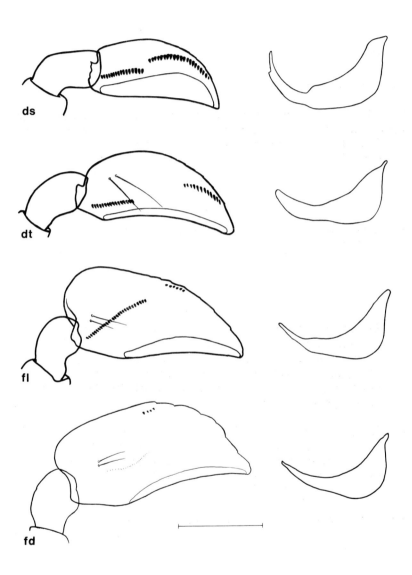

Fig. 50. Corixidae: Pala and right paramere of male *ds, Sigara dorsalis; dt, S. distincta; fl, S. falleni; fd, S. fallenoidea* (Scale line 0·5 mm).

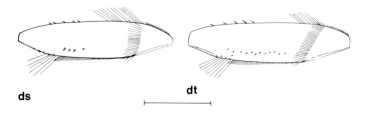

Fig. 51. Corixidae: Dorsal view of femur of posterior leg of *ds*, *Sigara dorsalis*; *dt*, *S. distincta* (Scale line 1 mm).

20   3–6 spines on the dorsal surface of the posterior femora (fig. 51*ds*); metasternal xiphus slightly concave with the tip pointing posteriorly (fig. 43*ds*); yellow transverse lines in anterior angle of clavus much wider than adjacent dark lines (fig. 49*ds*); strigil large and oval—

21

—   7–12 spines on the dorsal surface of the posterior femora (fig. 51*dt*); metasternal xiphus convex and curved so that the tip points dorsally (fig. 43*dt*); yellow transverse lines in anterior angle of clavus roughly equalling or narrower than dark lines (fig. 49*dt*, *fl*); strigil small and round— 22

The post-femoral spines are apparent only if correctly illuminated. A hind leg should be removed, taking care to note the dorsal surface, and then slowly rotated until the spines catch the light. Magnification ×40 is suitable.

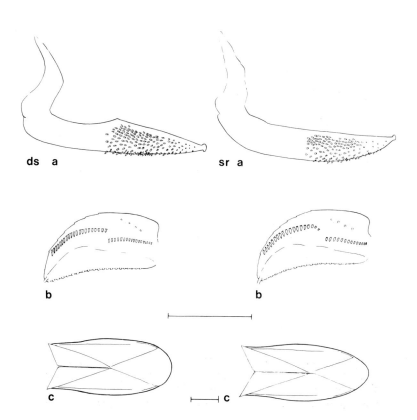

Fig. 52. Corixidae: *a*, left paramere; *b*, pala; *c*, outline of abdomen in dorsal view of male *ds*, *Sigara dorsalis*; *sr*, *S. striata*. (Upper scale line 0·5 mm; lower 1 mm).

21 Right paramere of male constricted abruptly before the apex (figs
    50*ds*, 53*ds a-f*); left paramere rising slightly to a point in the middle
    of the dorsal surface (fig. 52*a ds*); distal row of palar pegs not curving
    suddenly at their proximal end (fig. 52*b ds*); body broad and round
    posteriorly (fig. 52*c ds*).
    Length 6·5–8·0 mm—   **Sigara dorsalis** (Leach)

—   Right paramere tapering more or less uniformly to the apex (fig.
    53*sr h-l*); dorsal surface of left paramere flat (fig. 52*a sr*); distal row
    of palar pegs curving suddenly at their proximal end (fig. 52*b sr*);
    body narrower and more pointed posteriorly (fig. 52*c sr*).
    Length 6·5–8·0 mm—   **Sigara striata** (Linnaeus)

Sigara dorsalis is common and widespread while *S. striata* is confined to
parts of South East England (Kent and Sussex). Intermediate forms,
presumably hybrids, occur where their ranges overlap (figs 53*it g*, 70).

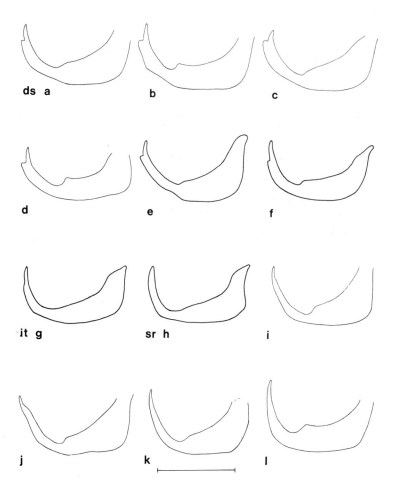

Fig. 53. Corixidae: Right paramere of male *ds a-f, Sigara dorsalis; it g*, intermediate between *S. dorsalis* and *S. striata; sr h-l, S. striata*. Specimens from *a*, Sweden; *b-d*, Lake District, England (Table 3, NR); *e*, Cheshire, England (MD); *f-h*, Kent, England (SE); *i-k*, Denmark; *l*, Germany. (Scale line 0·5 mm).

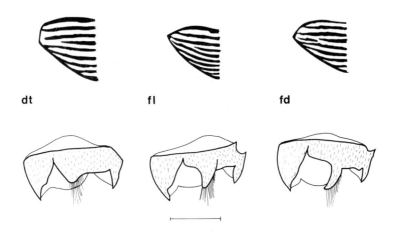

Fig. 54. Corixidae: Dorsal views of pronotum and seventh abdominal segment of male *dt*, *Sigara distincta*; *fl*, *S. falleni*; *fd*, *S. fallenoidea*. (Scale line 1 mm).

22(20) Male pala relatively long with the two rows of pegs approximately parallel with its longitudinal axis (fig. 50*dt*); central process on dorsal posterior margin of seventh abdominal segment of male short and rounded; anterior angles of pronotum obtuse (fig. 54*dt*); posterior margin of middle femur with a few long hairs (fig. 55*dt*).
Length 8·0–9·0 mm—   **Sigara distincta** (Fieber)

— Male pala very broad with the proximal row of pegs running diagonally (vestigial in *Sigara fallenoidea*) and the distal row placed near the dorsal margin (fig. 50*fl*, *fd*); central process on dorsal posterior margin of seventh abdominal segment of male relatively long; anterior angles of pronotum relatively acute (fig. 54*fl*, *fd*); posterior margin of middle femur with a fringe of long hairs along at least half its length (fig. 55*fl*, *fd*)—   **23**

The arrangement of hairs on the middle femur is a good diagnostic character but is difficult to see. A leg should be removed, placed in a drop of ethanol on a slide, and then rotated until the hairs catch the light.

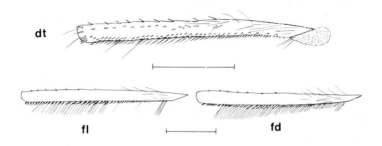

Fig. 55. Corixidae: Ventral view of femur of middle leg of *dt*, *Sigara distincta*; *fl*, *S. falleni*; *fd*, *S. fallenoidea*. (Scale lines 1 mm).

23   Proximal diagonal row of pegs on male pala very distinct (fig. 50*fl*); central process on dorsal posterior margin of seventh abdominal segment of male roughly triangular; anterior angles of pronotum markedly acute (fig. 54*fl*); posterior margin of middle femur with long hairs along distal half only (fig. 55*fl*).
Length 7·0–8·0 mm—   **Sigara falleni** (Fieber)

—   Proximal row of pegs on male pala vestigial (fig. 50*fd*); central process on dorsal posterior margin of seventh abdominal segment of male long, narrow and slightly hooked; anterior angles of pronotum intermediate between *S. distincta* and *S. falleni* (fig. 54*fd*); posterior margin of middle femur with long hairs along its entire length (fig. 55*fd*).
Length 7·0–8·0 mm—   **Sigara fallenoidea** (Hungerford)

Sigara distincta and *S. falleni* are widespread while *S. fallenoidea* is confined to Ireland.

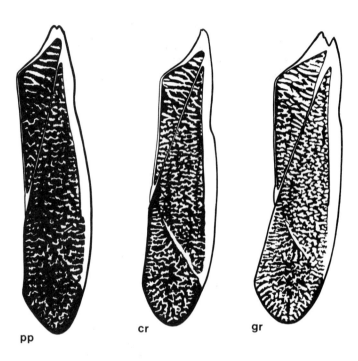

Fig. 56. Corixidae: Hemielytron of *pp*, *Glaenocorisa propinqua*; *cr*, *Arctocorisa carinata*; *gr*, *A. germari*. (Scale line 1 mm).

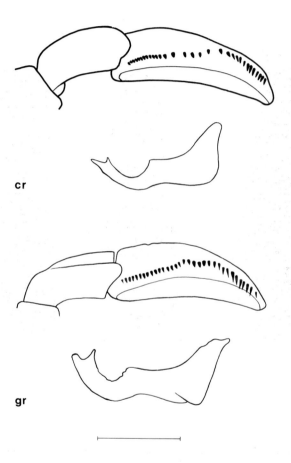

Fig. 57. Corixidae: Pala and right paramere of male *cr*, *Arctocorisa carinata*; *gr*, *A. germari*. (Scale line 0·5 mm).

24(19) Sides of metasternal xiphus virtually straight; anterior surface of
       male pala with an oblique ridge, distal palar pegs ending in bristles
       (fig. 58); posterior margin of the last complete abdominal sternum of
       female forming a regular uninterrupted curve (fig. 59*pp c*); long hairs
       on front of head forming a dense patch; females, as well as males,
       with a concavity on front of head; strigil <0·2 mm wide with fewer
       than six rows of teeth (fig. 59*pp a*); right paramere of male coming to
       a single point (fig. 58*pp*).
       Length 7·5–9·0 mm—   **Glaenocorisa propinqua** (Fieber) **25**

—   Sides of metasternal xiphus distinctly concave (fig. 43*gr*); anterior
       surface of male pala lacking an oblique ridge, palar pegs never ending
       in bristles (fig. 57); posterior margin of the last complete abdominal
       sternum of female with a central emargination (fig. 59*gr c*); long hairs
       on front of head sparse; concavity on front of head present only in
       males; strigil >0·2 mm wide with more than six rows of teeth (fig.
       59*gr a, cr a*); right paramere of male bifid (fig. 57)—   **26**

25   Bend in row of palar pegs reaching dorsal margin of male pala (fig.
       58*pp c*)—   **Glaenocorisa propinqua cavifrons** (Thomson)

—   Bend in row of palar pegs not reaching dorsal margin of male pala
       (fig. 58*pp p*)—   **Glaenocorisa propinqua propinqua** (Fieber)

I have seen specimens from the borders of England and Scotland which
are intermediate between these subspecies. Females cannot be separated.

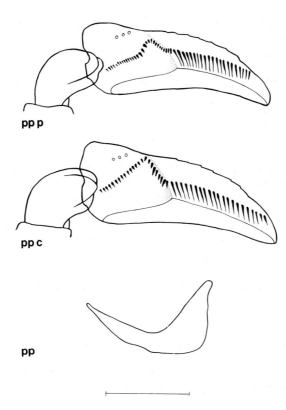

pp p

pp c

pp

Fig. 58. Corixidae: Pala of male *pp p*, *Glaenocorisa propinqua propinqua*; *pp c*, *G. propinqua cavifrons*. Right paramere of male *pp*, *G. propinqua*. (Scale line 0·5 mm).

26(24) Dark areas of hemielytra wider than light areas (fig. 56*cr*); posterior margin of middle femur of males beset along entire length with hairs whose length exceeds the width of the limb (fig. 59*cr b*); strigil about 0·25 mm wide with 7–9 rows of teeth (fig. 59*cr a*).
Length 8·0–10·0 mm—   **Arctocorisa carinata** (Sahlberg, C.)

— Light areas of hemielytra wider than dark areas (fig. 56*gr*); posterior margin of middle femur of males beset with hairs about half the width of the limb on its distal two-thirds, in addition a few long hairs (fig. 59*gr b*); strigil about 0·40 mm wide with 14–16 rows of teeth (fig. 59*gr a*).
Length 7·5–10·0 mm—   **Arctocorisa germari** (Fieber)

See note to couplet 22, p. 92.

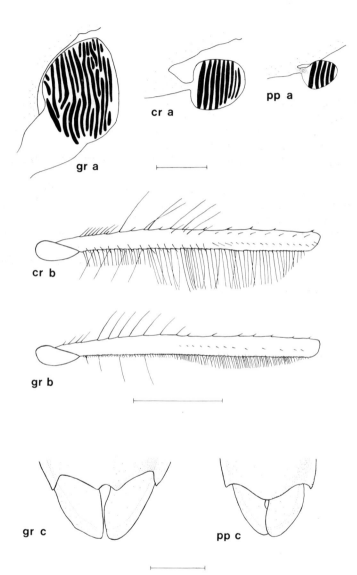

Fig. 59. Corixidae: Strigil of male *gr a, Arctocorisa germari; cr a, A. carinata; pp a, Glaenoco-risa propinqua.* (Scale line 0·25 mm). Ventral view of femur of middle leg of male *cr b, Arctocorisa carinata; gr b, A. germari.* Ventral view of the posterior end of abdomen of female *gr c, Arctocorisa germari; pp c, Glaenocorisa propinqua.* (Scale lines 1 mm).

27(18) Pronotum with 4–6 pale transverse lines—   **31**

—   Pronotum with more than 6 pale transverse lines—   **28**

28   Corium with longitudinal dark bands (fig. 60*lm, sm, vn*); pegs on
     male pala in two rows (fig. 61*lm, sm, vn*); concavity on front of male
     head deep and extending upwards between the eyes (fig. 62*lm, sm,
     vn*)—   **29**

—   Corium without longitudinal dark bands, or with an incipient band
     parallel with the claval suture (fig. 60*ng*); pegs on male pala in one row
     (fig. 61*ng*); concavity on front of male head shallow and terminated by
     a transverse ridge at the level of the lower margin of the eyes (fig.
     62*ng*); [general depth of colour very variable].
     Length 5·0–6·5 mm—   **Sigara nigrolineata** (Fieber)

29   Corium with two longitudinal dark bands, one near the inner, the
     other near the outer margin; sometimes an incipient band between
     the two (fig. 60*lm*); male pala with 12–16 pegs in distal row (fig.
     61*lm*); colour usually pale.
     Length 5·0–6·5 mm—   **Sigara limitata** (Fieber)

—   Corial markings as above but with the middle band usually complete
     (fig. 60*sm vn*); male pala with 4–7 pegs in distal row (fig. 61*sm, vn*);
     colour usually dark—   **30**

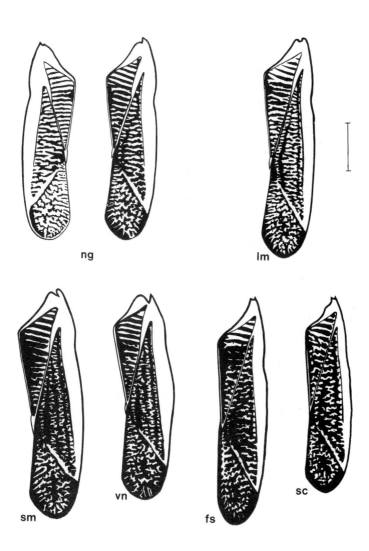

Fig. 60. Corixidae: Hemielytron of *ng*, *Sigara nigrolineata* (pale and dark specimens); *lm*, *S. limitata*; *sm*, *S. semistriata*; *vn*, *S. venusta*; *fs*, *S. fossarum*; *sc*, *S. scotti*. (Scale line 1 mm).

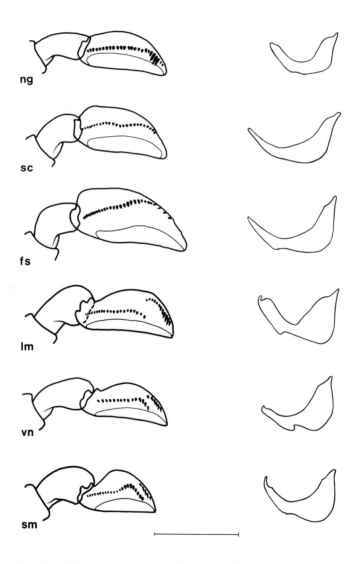

Fig. 61. Corixidae: Pala and right paramere of male *ng*, *Sigara nigrolineata*; *sc*, *S. scotti*; *fs*, *S. fossarum*; *lm*, *S. limitata*; *vn*, *S. venusta*; *sm*, *S. semistriata*. (Scale line 0·5 mm).

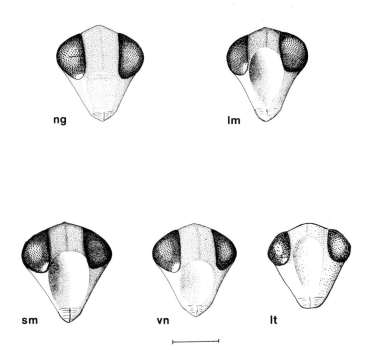

Fig. 62. Corixidae: Anterior view of head of male *ng, Sigara nigrolineata; lm, S. limitata; sm, S. semistriata; vn, S. venusta; lt, S. lateralis.* (Scale line 1 mm).

30   Metasternal xiphus longer than width at base (fig. 43*sm*); male pala
almost as wide as long (fig. 61*sm*); concavity on front of male head
extending upwards between the eyes (fig. 62*sm*).
Length 5·5–6·5 mm—   **Sigara semistriata** (Fieber)

—   Metasternal xiphus as long as width at base (as in fig. 43*ds*); male
pala distinctly less wide than long (fig. 61*vn*); concavity on front of
male head extending only just above the lower margin of the eyes
(fig. 62*vn*).
Length 4·5–6·0 mm—   **Sigara venusta** (Douglas & Scott)

31(27) Dark markings on hemielytra confluent in middle of clavus and
near inner angle of corium (fig. 60*fs*); dorsal edge of male pala forming
a sudden sharp curve just distal to its joint with the tibia, pegs crossing
the anterior surface diagonally (fig. 61*fs*).
Length 5·5-6·75 mm—   **Sigara fossarum** (Leach)

— Dark markings on hemielytra not confluent (fig. 60*sc*); dorsal edge of
male pala proceeding from its joint with the tibia in a smooth curve,
pegs forming an undulating line parallel with the longitudinal axis
(fig. 61*sc*).
Length 5·0-6·25 mm—   **Sigara scotti** (Douglas & Scott)

The confluent dark markings on the hemielytra of *Sigara fossarum* form
spots which are more easily seen with the naked eye rather than under
magnification.

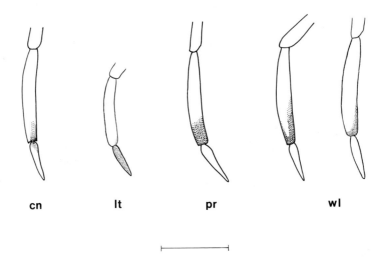

cn   lt   pr   wl

Fig. 63. Corixidae: Tarsus and claw of posterior leg of *cn, Sigara concinna*; *lt, S. lateralis*; *pr, Callicorixa praeusta*; *wl, C. wollastoni* (two forms). (Scale line 1 mm).

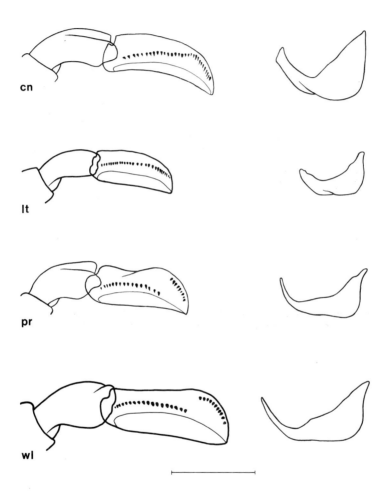

Fig. 64. Corixidae: Pala and right paramere of male *cn*, *Sigara concinna*; *lt*, *S. lateralis*; *pr*, *Callicorixa praeusta*; *wl*, *C. wollastoni*. (Scale line 0·5 mm).

32(11) Posterior claws entirely black or with a black spot at the base (fig.
63*cn*, *lt*; male pala with one row of pegs (fig. 64*cn*, *lt*); corial markings
irregular (fig. 65*cn*, *lt*)—   **33**

—   Posterior claws clear (fig. 63*pr*, *wl*); male pala with two rows of pegs
(fig. 64*pr*, *wl*); corial markings rather regular (fig. 65*pr*, *wl*)—   **34**

33 Posterior claw with a dark spot proximally and a matching spot on
the tarsus (fig. 63*cn*); pegs on male pala forming a relatively sharp
curve distally (fig. 64*cn*); concavity on front of male head shallow.
Length 6·0–7·5 mm—   **Sigara concinna** (Fieber)

—   Posterior claw entirely dark, the colour sometimes extending on to
the tarsus (fig. 63*lt*); pegs on male pala almost in a straight line (fig.
64*lt*); concavity on front of male head very deep (fig. 62*lt*).
Length 5·0–6·5 mm—   **Sigara lateralis** (Leach)

34(32) Dark mark on posterior tarsus usually square (fig. 63*pr*); male pala
with a central constriction on dorsal margin (fig. 64*pr*); hemielytral
markings on membrane distinct (fig. 65*pr*).
Length 7·0–8·0 mm—   **Callicorixa praeusta** (Fieber)

—   Dark mark extending along either side of distal end of tarsus with a
clear area between, occasionally restricted to one margin (fig. 63*wl*);
male pala without a constriction (fig. 64*wl*); hemielytral markings at
proximal margin of membrane showing little contrast between light
and dark areas (fig. 65*wl*).
Length 6·0–8·0 mm—   **Callicorixa wollastoni** (Douglas & Scott)

Both these species are variable in colour, related to the habitat from which
they originate. The degree of infuscation of the pale areas particularly
affects the degree of contrast between light and dark parts.

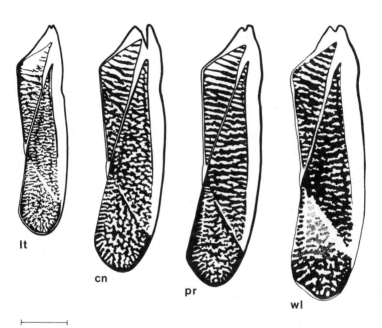

Fig. 65. Corixidae: Hemielytron of *lt, Sigara lateralis; cn, S. concinna; pr, Callicorixa praeusta; wl, C. wollastoni* (Scale line 1 mm).

35(3) Pronotum short with very short sides; head with a longitudinal
brown line just dorsal to each eye (fig. 66*sz a*); prestrigilar flap on
abdominal segment five of male as wide as long (fig. 66*sz d*); left
paramere twisted distally (fig. 66*sz b*); right paramere narrow and
ending in a smooth point (fig. 66*sz c*).
Length 2·0–2·5 mm—   **Micronecta scholtzi** (Fieber)

— Pronotum longer with longer sides; head lacking lines beside each
eye although it may have a central line (fig. 66*pw a*); prestrigilar flap
longer than wide (fig. 66*mn d, pw d*); left paramere not twisted (fig.
66*mn b, pw b*); right paramere ending in a point and also truncate
(fig. 66*mn c, pw c*)—   **36**

36 Hemielytron, immediately posterior to claval suture, generally light;
prestrigilar flap on abdominal segment five of male curved, posteriorly, towards mid-line of body (fig. 66*pw d*); right lateral margin
of seventh abdominal segment rather rounded (fig. 66*pw e*); truncate
tip of right paramere at right angles to longitudinal axis (fig. 66*pw c*).
Length 1·7–2·0 mm—   **Micronecta poweri** (Douglas & Scott)

— Hemielytron, immediately posterior to claval suture, generally dark;
prestrigilar flap straight and parallel-sided (fig. 66*mn d*); right lateral
margin of seventh abdominal segment produced into a distinct point
(fig. 66*mn e*); truncate tip of right paramere oblique (fig. 66*mn c*).
Length 1·7–2·2 mm—   **Micronecta minutissima** (Linnaeus)

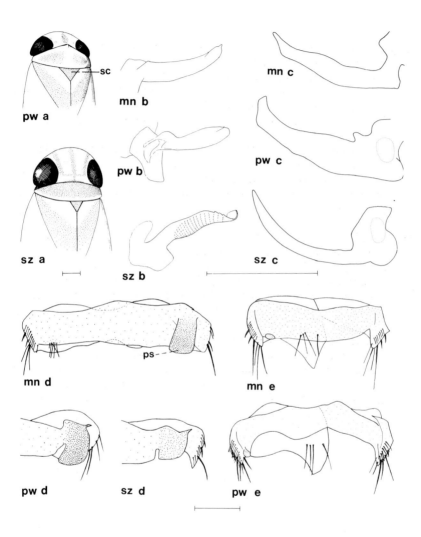

Fig. 66. Corixidae: Dorso-lateral view of anterior end of male *pw a, Micronecta poweri; sz a, M. scholtzi. (sc,* scutellum). Left and right parameres of male *mn b, mn c, M. minutissima; pw b, pw c, M. poweri; sz b, sz c, M. scholtzi.* Dorsal view of fifth abdominal segment of male *mn d, M. minutissima; pw d, M. poweri; sz d, M. scholtzi.* (*ps,* poststrigilar flap). Dorsal view of seventh abdominal segment of male *mn e, M. minutissima; pw e, M. poweri.* (Scale lines 0·25 mm).

ECOLOGY

INTRODUCTION

The study of ecology requires a knowledge and understanding of a complex suite of interrelationships. Furthermore, these interrelationships are unique to species, even their constituent populations and almost certainly to each individual. Thus, any attempt to provide a general account of a whole group of animals must, in spite of the need for objective standards, have idiosyncratic qualities. After the initial description of life cycles, the present account is based on two fundamental interpretations of the available evidence. Firstly, that a number of general correlations are demonstrable between the patterns of distributions of species and patterns of particular environmental factors. Secondly that there are a number of categories of information related to causal processes which appear to be of more fundamental importance than others; these are placed under separate sub-headings. However, a too rigid separation of the subject matter is undesirable since such a procedure would fail to reveal important inter-relationships.

LIFE CYCLES

The Gerromorpha and Nepomorpha have life cycles consisting of an egg stage, five nymphal (larval) instars, and the adult instar. The nymphs are essentially miniature adults both in their morphology and ecological requirements, as would be expected in members of the Hemimetabola. The annual cycle of the British species consists of either one (univoltine), or apparently two (bivoltine) generations. However, there is evidence which suggests that any second generation is incomplete (see section on wing polymorphism, p. 150). Casual observations suggest the possibility of three generations in a few species but I know of no firm evidence based on thorough studies of life cycles. Reproduction occurs during the summer and all, except Micronectinae, overwinter as adults. The patterns of nymphal development in four representative species, from three families, two univoltine and two apparently bivoltine, indicate the range of sequences in time that are typical of the entire group (fig. 67).

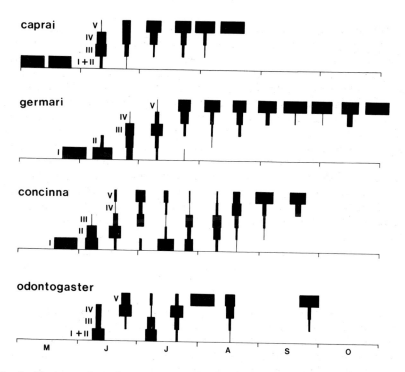

Fig. 67. The proportions of nymphal instars (I-V) during the reproductive season of *Velia caprai* (data from Brinkhurst 1959a); *Arctocorisa germari* (data from Crisp 1962b); *Sigara concinna* (data from Savage 1971a); *Gerris odontogaster* (data from Brinkhurst 1959a).

Mating behaviour

In Corixidae the ovaries begin to develop at various times between February and March (Young 1965b). Mating first takes place in April–May in most species of Veliidae, Gerridae, Notonectidae and Corixidae (e.g. Brinkhurst 1966; Crisp 1962b; Savage 1971a,c; Southwood & Leston 1959). Mating behaviour has been studied thoroughly in Corixinae and Gerridae. In Corixinae the sexes aggregate during April in a series of dense clusters near the lake shore, with the majority paired (Crisp 1962b; Savage 1971a). This aggregation is accompanied by intense stridulation (Aiken 1985). Both male and female Corixinae stridulate by rubbing fields of small pegs (the pars stridens), on the fore femora, over thickened flanges (the plectrum) on the maxillary plate of the head which, in turn, is almost certainly the sound-producing organ (Aiken 1982, 1985; Jansson 1972). In *Palmacorixa nana* Walley, a North American species, females showed a significant positive exponential response towards the mean call rate of the densest male aggregation while males showed a preference for the less dense aggregations. At high densities males showed a higher number of attempts at copulation but suffered greater interference from other males. Thus, the differing responses of males and females tend to ensure that costs and benefits in mating success are balanced (Aiken 1982). Males of *P. nana* produce four distinct acoustic signals: a spontaneous call, a courtship call, a mounting call and a copulatory call. The latter two are produced by a different mechanism from the former, namely, by drawing a row of pegs on the inner face of the middle femora across the edges of the hemielytra. In contrast, females produce only a single agreement call (Aiken 1982). In the genus *Cenocorixa* only the first two male calls and the female call have been reported (Jansson 1973, 1976). The precise function and need for these call signals in mating varies in different species. However, it is possible to elucidate some general principles. The spontaneous male call is concerned with the aggregation of the sexes. The male then produces a courtship call which elicits the agreement call in a receptive female. The female remains stationary while the male, oriented by the agreement call, swims directly towards her. In *P. nana* mounting and copulation ensue accompanied by the respective calls; other species remain silent (Aiken 1982; Jansson 1973). Although attention has been directed at the intense period of reproductive activity in spring, it should be emphasised that mating continues throughout the summer.

An important distinction between species lies in the precise acoustic characteristics of their stridulatory signals, a phenomenon noted in the British corixid fauna (Leston 1955). These signals clearly play an important part in ensuring conspecific mating in sympatric species which, in turn, contributes to more or less temporary habitat partitioning. Audiospectrographic analysis of stridulation in two pairs of closely related species, *Sigara striata* and *S. dorsalis* on the one hand and *Arctocorisa carinata* and *A. germari* on the other, has shown that they have distinct but related patterns respectively. Similar distinctions were found in the geographically isolated montane populations of *A. carinata* (Jansson 1979a, 1979b, 1979c; Jansson & Pajunen 1978). Elsewhere, it has been argued that the morphology of the male palae is peculiar to each species and related to mating and feeding (Popham *et al.* 1984). Thus, these features collectively form a suite of morphological and behavioural characteristics associated with the adaptive radiation and ecological distribution of the Corixinae.

In Gerridae, sex discrimination prior to mating appears to be dependent upon ripple signals produced on the water surface by vertical movements of the legs and abdomen. Males of *Limnoporus rufoscutellatus* produce three types of ripple signal: a high frequency (*c.* 25 Hz) repel signal, a low frequency (*c.* 3 Hz) courtship signal, and a medium frequency (*c.* 10 Hz) threat signal. Males produce repel signals when approached by another individual. If a repel signal is not produced in response, the male judges that the individual approaching is female (not always correctly!) and emits the courtship signal. A receptive female responds by lowering the abdomen and allowing the male to mount. Unreceptive females raise the abdomen and may emit a ripple signal. Male–male encounters are resolved by immediate attack and retreat. Some males are territorial: they anchor themselves to a convenient object, such as a floating attached leaf, and defend a territory of approximately 50 cm diameter by successfully attacking intruding males. In contrast, they court approaching females. Territorial males have greater mating success than vagrant males (Vepsäläinen & Nummelin 1985b). Similar behaviour patterns occur in North American members of the genus *Limnoporus* (Spence & Wilcox 1986; Wilcox & Spence 1986). Females of *Gerris najas* are territorial in relation to food resources rather than as a mating strategy. Males remain mounted on the female for more or less the whole reproductive season, even when the female dives to lay eggs (Vepsäläinen & Nummelin 1985a). This prolonged period of mounting may be related to similar behaviour in *Gerris lateralis*, where it has been shown that the last male to mate fertilizes 80% of eggs subsequently laid. Thus prolonged post-copulatory mounting is an advantage to individual males, by ensuring paternity (Arnqvist 1988).

Oviposition

Ovipositing females of water bugs usually attach their eggs to an underwater surface although a few simply deposit them among plants at the water edge (Southwood & Leston 1959). Furthermore, there are characteristic differences, often between closely related species, which suggest that the choice of oviposition site is an adaptive strategy of importance to the distribution of species. *Notonecta maculata*, *Micronecta poweri* and *Arctocorisa germari* lay eggs exclusively on stones, in nature (Crisp 1962b; Southwood & Leston 1959; Walton 1936). All are associated with habitats which contain little vegetation (see Table 3). In contrast, *Notonecta glauca* and *Sigara concinna* lay eggs on plants while *Corixa punctata*, *Sigara dorsalis*, *S. falleni*, *S. stagnalis*, *Hesperocorixa castanea* and *S. lateralis* lay on a wide variety of materials, although the last named species prefers stones (Leston 1955; Savage 1979b). The numbers of eggs laid by individual female Corixidae varies from 10 to 1000 (Peters & Spurgeon 1971; Young 1965b) and eggs may be laid at different times of the year in different species, in relation to the temperature of a particular habitat (Jansson & Scudder 1974). In the British species *Sigara dorsalis* and *S. scotti*, individual females laid 300–400 eggs when supplied with abundant food under laboratory conditions (Young 1965b); overwintered females of *S. concinna*, *S. lateralis* and *S. stagnalis* each laid only 13–20 eggs while *S. lateralis* females of the second generation laid 30–40 eggs in 14 days under similar conditions (Savage 1979b). The mean number of eggs per female during the whole oviposition season was 213 in *Arctocorisa germari* (Crisp 1962a). The total oviposition periods were similar in the univoltine *A. germari* (25 April–28 August) and the partially bivoltine species, *S. concinna* (5 May–8 June; 20 July–25 August) and *S. lateralis* (27 April–23 September, 1974 and 22 April–30 May; 15 July–2 September, 1975) (Crisp 1962a; Savage 1979b).

Kaitala (1987) kept ovipositing females of *Gerris thoracicus* and *G. lacustris* at constant temperature but provided a series of groups with different amounts of food. In all experiments, females laid more eggs when food was superabundant than when it was scarce, 100–350 and 20–100 eggs female^{-1} respectively. Furthermore, females of *G. thoracicus* showed a rapid response to temporal variations in food supply, with very high oviposition rates when food was superabundant and virtual cessation of oviposition when food was scarce. There was a significant difference in the longevity of females of the two species in relation to the availability of food. *G. thoracicus* lived for 17–24 days with superabundant food, and for 28–40 days with little food. In contrast the comparable survival times for *G. lacustris* were 32 and 20 days. Thus there was an inverse relationship. The longevity of males remained constant in relation to food supply.

G. *thoracicus* is a species of temporary habitats whereas G. *lacustris* is found in rather more permanent places (cf. Table 5, p. 151). G. *thoracicus* has evolved an adaptive strategy which enables it to make a dynamic response to the rapidly changing conditions of temporary habitats whereas G. *lacustris* lacks these dynamic features (Vepsäläinen *et al.* 1985).

Development

The rates of development of eggs and nymphs have been estimated in relation to temperature, in *Arctocorisa germari*, *A. carinata* and *Callicorixa producta* (Reuter) among Corixidae (Crisp 1962b; Pajunen & Sundbäck 1973). Similar studies of total development time, from egg to adult, have been made on some North American Gerridae (Spence *et al.* 1980). Regression analysis of the data shows that the best fit is with a power equation, logarithmically transformed: $\ln \bar{D} = \ln a + b \ln T$, where $\bar{D}$ is mean duration (days) of the relevant part of the life cycle, T is temperature (°C) and a and b are constants. Pajunen & Sundbäck (1973) found that their laboratory experiments enabled accurate prediction of development time in the field and that *A. carinata*, a glacial relict species, was more resistant to low temperature than *C. producta*. There is such a close agreement between the stages and species of Corixidae and Gerridae, in relation to development and temperature, it is tempting to predict that a standard pattern occurs in water bugs (fig. 68b; Table 2). The patterns of growth of the nymphal instars of *Sigara concinna*, *Arctocorisa germari* and *Corixa panzeri*, (I), as reflected by an increase in mean head or body width ($\bar{W}$), fit the equation: $\ln \bar{W} = a + bI$, where a and b are constants. Again, the three species show remarkable consistency (fig. 68a; Table 2) (Crisp 1962b; Savage 1971b; Sutton 1947).

TABLE 2. REGRESSION EQUATIONS AND CORRELATION COEFFICIENTS (r) FOR THE GROWTH RELATIONSHIPS SHOWN IN FIG. 68*

Species and structure-stage	Regression equation	r
Sigara concinna head width	$\ln Y = 0\cdot260X - 0\cdot641$	$0\cdot995$
Arctocorisa germari head width	$\ln Y = 0\cdot258X - 0\cdot460$	$0\cdot998$
Corixa panzeri body width	$\ln Y = 0\cdot298X - 0\cdot116$	$0\cdot970$
Arctocorisa carinata egg	$\ln Y = 8\cdot32 - 1\cdot99 \ln X$	$-0\cdot990$
Arctocorisa carinata 2nd instar nymph	$\ln Y = 7\cdot06 - 1\cdot77 \ln X$	$-0\cdot980$
Arctocorisa carinata 5th instar nymph	$\ln Y = 6\cdot95 - 1\cdot50 \ln X$	$-0\cdot993$
Gerris pingreensis egg to adult	$\ln Y = 8\cdot54 - 1\cdot64 \ln X$	$-0\cdot999$
Gerris comatus egg to adult	$\ln Y = 9\cdot68 - 1\cdot96 \ln X$	$-0\cdot978$

* $\ln$ = natural logarithms $(\log_e)$

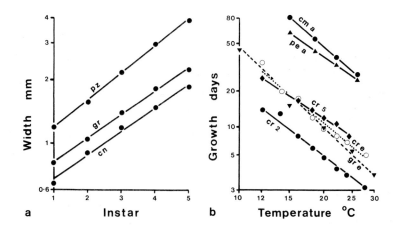

Fig. 68. a, The relationship between mean head width or mean body width (Y mm) and instar number (X) in cn, *Sigara concinna* – head width (data from Savage 1971b); gr, *Arctocorisa germari* – head width (data from Crisp 1962b); pz, *Corixa panzeri* – body width (data from Sutton 1947). b, The relationship between mean growth (Y days to complete a particular life history stage) and temperature (X °C) in cr e, *Arctocorisa carinata* – egg; cr 2, A. *carinata* – 2nd instar; cr 5, A. *carinata* – 5th instar (data from Pajunen & Sundbäck 1973); gr e, A. *germari* – egg (data from Crisp 1962b); pe a, *Gerris pingreensis* Drake & Hottes – egg to adult; cm a, G. *comatus* Drake & Hottes – egg to adult (data from Spence et al. 1980). See Table 2 for details of the regression equations.

Population dynamics

The seasonal population dynamics of adult Corixidae conform to a general pattern in relatively stable communities. Numbers reach a distinct maximum in autumn with a suggestion of a minor maximum in early summer (fig. 69) (Crisp 1962a; Savage 1979a; Walton 1943). This pattern did not occur in unstable rock pool habitats in Finland among adults, although nymphs showed a similar bimodal distribution (Pajunen 1977). Also, seasonal population dynamics were very varied, due mainly to immigration and emigration, in a saline lake in Cheshire when environmental conditions and corixid communities were changing (Savage 1979a, 1981).

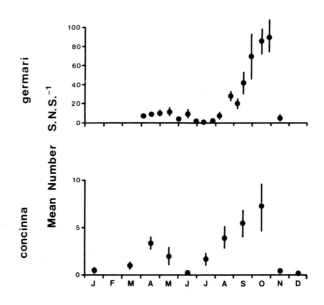

Fig. 69. The mean numbers of adults per standard net sweep (±95% confidence limits) for *Arctocorisa germari* at a station in Barbrook reservoir, in 1957 (data from Crisp 1962a), and for *Sigara concinna* at a station in Watch Lane Flash, in 1970 (data from Savage 1979a). Note that each author used his own individual type of standard net sweep (SNS).

Production

Crisp (1962a) attempted to estimate production in *Arctocorisa germari* in an upland reservoir. During the reproductive season of 1958, $4\,\mathrm{g\,m}^{-2}$ wet weight of eggs were produced of which $2\,\mathrm{g\,m}^{-2}$ were lost; the final adult biomass was $48\,\mathrm{g\,m}^{-2}$ wet weight and $71\,\mathrm{g\,m}^{-2}$ were lost during nymphal development. Thus production was estimated as $121\,\mathrm{g\,m}^{-2}$ wet weight $(2+71+48)$, or $29\cdot4\,\mathrm{g\,m}^{-2}$ dry weight. The use of a graphical method for estimating production (Allen 1951) from Crisp's data, gives values of $138\,\mathrm{g\,m}^{-2}$ wet weight and $33\cdot6\,\mathrm{g\,m}^{-2}$ dry weight.

GEOGRAPHICAL AND ECOLOGICAL DISTRIBUTION

Present knowledge of the geographical and ecological distribution of probable breeding populations of British aquatic Hemiptera Heteroptera is summarised in cartographic and tabular form (figs 70, 71, 72; Table 3). Little need be added concerning geographical distribution, since well supported causal explanations are not available. However, two pairs of species are of particular interest. *Corixa punctata* and *C. iberica* appear to be virtually allopatric, with the latter confined to the extreme North West of the British Isles (fig. 71). These species are closely related but I have not seen any specimens which may be described as intermediate in form although some *C. punctata* from the West of Mainland Scotland (SM, fig. 72) have less well-developed central processes on their right parameres than those from the Midlands (MD, fig. 72). *C. punctata* is found throughout Europe while *C. iberica* is recorded only from south west Spain and Portugal (Jansson 1986). *Sigara dorsalis* and *S. striata* are allopatric except in the extreme South East where intermediate forms occur which must be hybrids (figs 52, 53, 70; Table 3). These two species form part of a complex of patterns of speciation occurring in Europe (Jansson 1986).

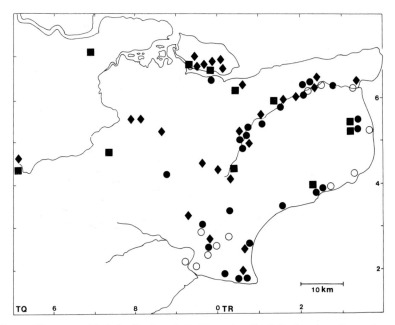

Fig. 70. The geographical distribution of: ■, *Sigara dorsalis*; ● ○, *S. striata*; ◆. specimens
intermediate between the two, in Kent and Sussex (South East quarter of region SE
in fig. 72). Solid symbols, specimens seen by A.A.S.; open symbols from Lansbury &
Leston 1966. Specimens examined from Maidstone Museum & Art Gallery.

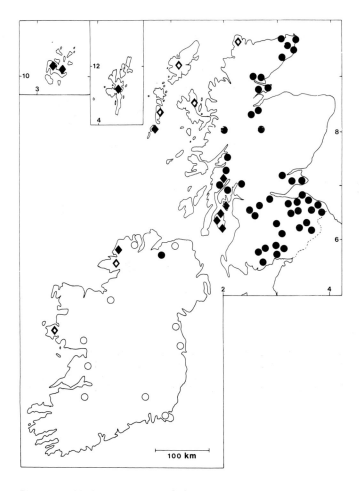

Fig. 71. The geographical distribution of: ● ○, *Corixa punctata*, and ◆ ◇, *C. iberica*, in Scotland and Ireland (regions SM, SI and IR in Fig. 72). Solid symbols, specimens seen by A.A.S.; open symbols from Jansson 1986. Specimens examined from Hope Entomological Collections, University Museum, Oxford; National Museums & Galleries on Merseyside, Liverpool; Museum & Art Gallery, Perth; Royal Museum of Scotland, Edinburgh; Museum & Art Gallery, Inverness; Caithness Biological Records Centre, Wick.

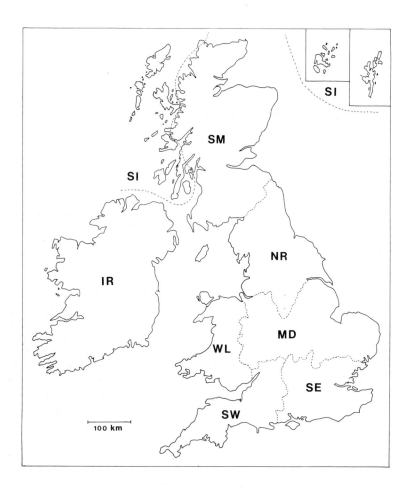

Fig. 72. Map of the British Isles showing the geographical distribution regions listed in Table 3. SW, South West; SE, South East; MD, Midlands; WL, Wales; NR, North England; SM, Scotland – Mainland; SI, Scotland – Western Isles with Orkney and Shetland; IR, Ireland. The division into regions is based on national and county boundaries prior to 1974 with the following exceptions: only the upland parts (*c.* > 300 m) of Derbyshire and Staffordshire are included in NR; the division between SM and SI follows Conrad's Index of Continentality (Chandler & Gregory 1976).

TABLE 3. THE GEOGRAPHICAL AND ECOLOGICAL DISTRIBUTION OF BRITISH AQUATIC HEMIPTERA HETEROPTERA

The main body of the table appears on pages 126–129.

Geographical Distribution

The British Isles are divided into eight regions as in fig. 72. The four circles each represent a percentage frequency of a given species in relation to the total catch recorded for that region:

● >10%;　● 9–5%;　● 4–2%;　• <2%

Ecological Distribution

Environmental factors are divided into five main categories on a qualitative basis; each is then subdivided quantitatively. The relative sizes of the four circles within a main category give an indication of the relative distribution of a species within that category (i.e. type of water body, habitat, conductivity, pH or altitude).

● common;　● frequent;　● occasional;　• rare;　Ⓢ Scotland only.

The compilation for the geographical distribution of species is based on sources which provide quantitative data rather than isolated records. Where there is lack of evidence of a breeding population, records are ignored. Quantitative data are interpreted as: (1) actual counts of individual specimens; (2) records of the number of habitats containing a species out of a total number of habitats studied; (3) the number of areas, usually tetrads, in which a species is recorded out of the total number in which collections were made. Regression analysis showed a significant positive correlation between (1) and (2)+(3): $\ln Y = a + b \ln X$, where Y is percentage habitats (2) or areas (3) and X is percentage numbers (1) for a given species in a given region; a and b are constants. Thus values of (2) and (3) were converted to values of (1) from the equation: $\ln Y = 1\cdot71 + 0\cdot575 \ln X$; $r = 0\cdot90$, $p<0\cdot001$, df $= 40$. A range of transformations of data was tried. They agreed closely with the simple method described above.

A similar treatment of ecological data was rejected owing to the lack of suitable evidence and hence this part of the table is more subjective.

References: Brinkhurst 1959b; Brown 1943, 1948; Carrick & Sutcliffe 1982; Gillespie 1985; Groves 1982; Jansson 1977a, 1977b, 1987; Jastrey 1981; Kirby 1983a, 1983b; Lansbury 1954, 1965; Leston 1955, 1958; Macan 1938, 1949, 1954a, 1957, 1970; Macan & MacFadyen 1941; McCarthy & Walton 1980; Martin 1970; Morris 1969; Nau 1979; Pearce & Walton 1939; Popham 1943, 1949, 1950; Read 1987; Savage 1971c, 1979a, 1981, 1982b; Savage & Pratt 1976; Southwood & Leston 1959; Vepsäläinen 1973b; Walton 1942, 1943, 1981.

SPECIES	Geographical Region								Water Body Type			Habitat Type
												Increasing plant cover OR Organic matter →
	SE	SW	MD	WL	NR	SM	SI	IR	River	Lake	Pond or Pool	
Mesovelia furcata	●	●	●	●							●	●
Hebrus pusillus	●	●	●							●	●	●
Hebrus ruficeps	●	●	●	●			●		●	●	●	●
Hydrometra gracilenta	●		●								●	
Hydrometra stagnorum	●	●	●	●	●			●	●	●	●	
Velia caprai	●	●	●	●	●	●	●		●	●	●	●
Velia saulii	●			●	●	●		●	●	●		
Microvelia pygmaea	●	●								●		●
Microvelia reticulata	●	●	●	●			●			●	●	●
Microvelia buenoi umbricola			●							●	●	●
Gerris costai poissoni				●	●	●	●	●			●	
Gerris lateralis asper			●	●	●	●	●				●	●
Gerris thoracicus	●	●	●						●	●		●
Gerris gibbifer	●	●	●	●							●	●
Gerris argentatus	●	●								●	●	●
Gerris lacustris	●	●	●	●	●	●	●		●	●	●	
Gerris odontogaster	●	●	●	●	●	●				●	●	●
Gerris najas	●	●							●	●		●
Gerris paludum	●		●						●		●	
Limnoporus rufoscutellatus								●	●	●		
Nepa cinerea	●	●	●		●			●		●	●	●
Ranatra linearis	●	●	●							●	●	
Ilyocoris cimicoides	●	●	●	●						●	●	●
Aphelocheirus aestivalis	●	●	●	●	●			●	●			●
Notonecta glauca	●	●	●	●	●	●		●		●	●	●
Notonecta marmorea viridis	●	●									●	●
Notonecta obliqua	●	●	●		●			●		●	●	●
Notonecta maculata	●	●	●					●			●	●
Plea leachi	●	●	●		●			●	●	●		●
Micronecta scholtzi	●	●							●	●		●
Micronecta minutissima	●								●	●		●
Micronecta poweri	●	●	●	●	●	●		●	●	●		●

Conductivity μS cm⁻¹ k 25				pH		Altitude m		Species
<100	100–1000	1000–10,000	>10,000	<6	>6	<300	>300	
						●		Mesovelia furcata
						•		Hebrus pusillus
●				●		•	•	Hebrus ruficeps
						●		Hydrometra gracilenta
●	●				•	●		Hydrometra stagnorum
●	●			•	•	●	•	Velia caprai
●	●			●	•	●		Velia saulii
•	•					●		Microvelia pygmaea
•	•	•		•	•	●		Microvelia reticulata
	•				•	●		Microvelia buenoi umbricola
•				●		•	●	Gerris costai poissoni
								Gerris lateralis asper
	•	•			●	●		Gerris thoracicus
●	•			●		●	•	Gerris gibbifer
•	•				•	•		Gerris argentatus
●	●	•	•	●	●	●	•	Gerris lacustris
•	●	•	•	●	•	●	•	Gerris odontogaster
●	•				●	●		Gerris najas
					●	●		Gerris paludum
						●		Limnoporus rufoscutellatus
•	●	•		•	●	●		Nepa cinerea
						●		Ranatra linearis
●	•				●	●		Ilyocoris cimicoides
	•				●	●		Aphelocheirus aestivalis
•	●	●		•	●	●		Notonecta glauca
	•	●			●	●		Notonecta marmorea viridis
●	•			●	•	●	●	Notonecta obliqua
	•				•	●		Notonecta maculata
●	●			•	•	●	•	Plea leachi
						●		Micronecta scholtzi
	•			•		•		Micronecta minutissima
●	●			•		●	•	Micronecta poweri

SPECIES	Geographical Region								Water Body Type			Habitat Type
	SE	SW	MD	WL	NR	SM	SI	IR	River	Lake	Pond or Pool	Increasing plant cover OR Organic matter →
Cymatia bonsdorffii	•	•	•	●	●	•	●	●		●	●	• ●
Cymatia coleoptrata	●	•	●							●	●	: ●
Glaenocorisa p. propinqua		•		•	•	●		•		●		● ●
Glaenocorisa p. cavifrons				•	•	●				●		● ●
Callicorixa praeusta	●	•	●	●	●	●	•	●		●	●	● ●
Callicorixa wollastoni			•	●	●	●	●			•	●	• ●
Corixa dentipes	•	•	•	•	•			•		•	●	• •
Corixa punctata	●	●	●	●	•	•		•		•	●	• ●
Corixa iberica							●			•	●	
Corixa affinis	•	•	•							•	•	•
Corixa panzeri	•	•	•			•					●	•
Hesperocorixa linnaei	●	•	•	•	●	●	•	•		●	●	● •
Hesperocorixa sahlbergi	●	●	●	●	●	●	•	•		●	●	• ●
Hesperocorixa castanea	•	•	●	●	•	•		•		●	●	• ●
Hesperocorixa moesta	•	•	•	•				•		•		• •
Arctocorisa carinata					●	•	●			●		●
Arctocorisa germari	•	•	•	•	•	•	•	•		●		●
Sigara dorsalis	●	●	●	●	●	●	•	●	•	●		● •
Sigara striata	●	●	●	●	●	●	•	•	●	•	•	• ●
Sigara distincta	●	●	•	●	●	●	●	•		●	•	• ●
Sigara falleni	●	●	●	●	●	●	•	•	•	●	•	• ●
Sigara fallenoidea											•	●
Sigara fossarum	•	●	•	•	•	•	•		●		•	• •
Sigara scotti			•	●	●	●	●	•				● •
Sigara lateralis	●	●	•	●	•	•		•		•	●	• ●
Sigara nigrolineata	●	●	•	●	•	•	•	•		•	●	• ●
Sigara concinna	•	•	•	•	•	•		•		•	●	• •
Sigara limitata	•	•	•	•	•	•				•	•	• •
Sigara semistriata	•	•	•	•	•	•		●		•	•	• ●
Sigara venusta	•	•	•	●	●	●	•		●	•	•	• ●
Sigara selecta	•	•	•					•			●	• ●
Sigara stagnalis	•	•	●	•	•			•			●	● ●

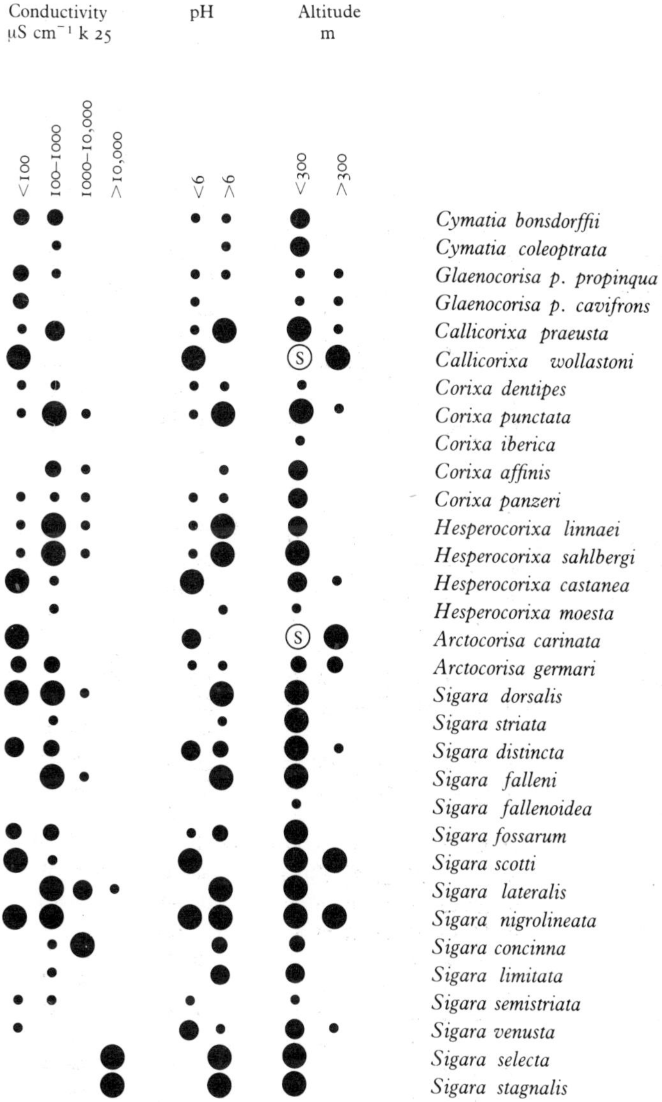

Conductivity pH Altitude
µS cm⁻¹ k 25 m

Conductivity µS cm⁻¹ k 25				pH		Altitude m		
<100	100–1000	1000–10,000	>10,000	<6	>6	<300	>300	

Cymatia bonsdorffii
Cymatia coleoptrata
Glaenocorisa p. propinqua
Glaenocorisa p. cavifrons
Callicorixa praeusta
Callicorixa wollastoni
Corixa dentipes
Corixa punctata
Corixa iberica
Corixa affinis
Corixa panzeri
Hesperocorixa linnaei
Hesperocorixa sahlbergi
Hesperocorixa castanea
Hesperocorixa moesta
Arctocorisa carinata
Arctocorisa germari
Sigara dorsalis
Sigara striata
Sigara distincta
Sigara falleni
Sigara fallenoidea
Sigara fossarum
Sigara scotti
Sigara lateralis
Sigara nigrolineata
Sigara concinna
Sigara limitata
Sigara semistriata
Sigara venusta
Sigara selecta
Sigara stagnalis

Differences in the ecological distribution of species occur both *between* different types of water body and *within* each type; the latter is often referred to as habitat partitioning. A water body may be defined in terms of three fundamental sets of abiotic components. These are size–shape, water chemistry and whether still (lentic) or flowing (lotic). These components are determined by the climate and geology of the catchment which, in turn, influence biotic and anthropogenic effects. The majority of water bugs inhabit lentic rather than lotic waters (Table 3). Lentic waters are particularly subject to processes of terrestrialization initiated by inorganic and organic silting, followed by the development of vegetation and the consequent accumulation of further organic matter (e.g. Macan 1970). Two other factors, primarily related to size–shape, are important, namely, shelter and habitat stability. The attention of ecologists, in relation to these factors, has been concentrated on the two families of water bugs, Corixidae and Gerridae, which show the highest species diversity. Particular attention has been paid to Corixidae in the British Isles while Gerridae have been subject to similar studies in Scandinavia and North America. Studies of Notonectidae have been added in recent years.

The Corixidae are divisible into species typical of rivers, lakes and ponds respectively (Macan 1954a). They are further divisible, in lakes and ponds, according to the water chemistry of the habitat (Jansson 1977a, 1977b, 1987; Macan 1954a, 1955a; Savage 1971c, 1982b; Scudder 1976). The electrical conductivity of waters shows a significant correlation with ionic composition, both quantitatively and qualitatively, and may be used as a general indicator in freshwater and oligohaline habitats (Savage 1977, 1981, 1982a, 1982b). Accordingly, the distribution of some lake species may be defined (fig. 73). Studies of the effects of the size of a water body on water bug communities, apart from those of Macan (1954a), have been neglected until recently, perhaps owing to the unstable nature of small habitats and the consequent effects (Bröring & Niedringhaus 1988). However, it is clear that the distribution of *Sigara dorsalis* and *S. falleni* in eutrophic lakes and ponds ($300-1000\,\mu S\,cm^{-1}\,k\,25$) is related to a function of size–shape (fig. 74) (Savage 1982b).

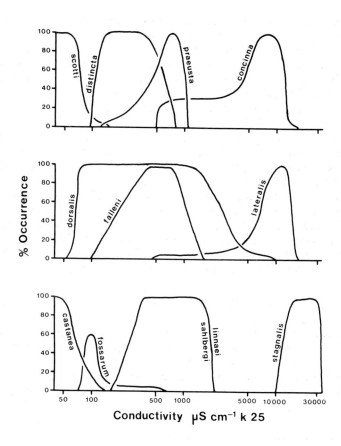

Fig. 73. A semi-schematic representation of the relationship between the maximum probable occurrence (Y, %) of some species of Corixidae, and conductivity (X, μS cm^{-1} k 25) in lakes. (Data from Carrick & Sutcliffe 1982; Macan 1938, 1949, 1954a, 1970; Martin 1970; Savage 1971c, 1979a, 1981, 1982b; Savage & Pratt 1976; Walton 1943).

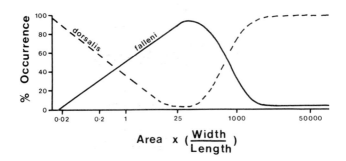

Fig. 74. A semi-schematic representation of the relationship between the maximum probable occurrence (Y, %) of *Sigara falleni* and *S. dorsalis*, with the 'Shelter Factor' (X, area in hectares × width/length), in lakes and ponds where the overall conductivity range is 300–1000 µS cm^{-1} k 25. (Data from Savage 1982b).

The species succession of Corixidae in relation to the presence of vegetation and the accumulation of organic matter was first demonstrated (Macan 1938) in the essentially oligotrophic Cumbrian lakes (40–110 µS cm^{-1} k25; Carrick & Sutcliffe 1982). The succession (fig. 75) is:

$$poweri \rightarrow dorsalis \begin{array}{c} \nearrow \quad fossarum \quad \searrow \\ \searrow \quad distincta \quad \nearrow \end{array} scotti \rightarrow castanea.$$

In the eutrophic lakes of the North West Midlands (300–900 µS cm^{-1} k25) (Macan 1967; Savage & Pratt 1976; Savage in preparation) the succession (fig. 76) is:

$$falleni \rightarrow praeusta \rightarrow linnaei \rightarrow sahlbergi.$$

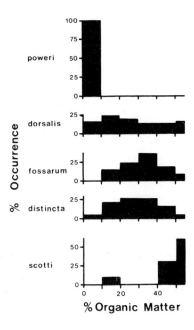

Fig. 75. The occurrence (Y, %) of some species of Corixidae in relation to organic matter (X, %) in the substratum of Windermere. (Data from Macan 1970).

Communities with intermediate characteristics occur in more meso-trophic lakes (100–300 μS cm^{-1} k 25) while in very slightly oligohaline lakes (2000–4000 μS cm^{-1} k 25) $S.$ $dorsalis$ replaces $S.$ $falleni$ and is itself replaced by $S.$ $concinna$ (4000–9000 μS cm^{-1} k 25) which, in turn, is re-placed by $S.$ $stagnalis$ (9000–30,000 μS cm^{-1} k 25). $S.$ $lateralis$ tends to replace $S.$ $concinna$ if organic matter accumulates (Savage 1971c, 1979a, 1981, 1985). Other studies are in general agreement with these associations (Bröring & Niedringhaus 1988; Popham 1949; Walton 1943).

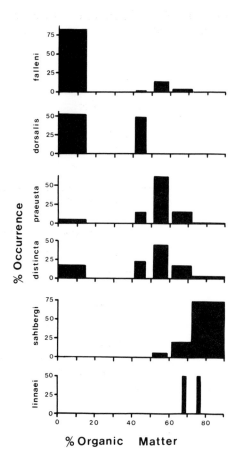

Fig. 76. The occurrence (Y, %) of some species of Corixidae in relation to organic matter (X, %) in the substrata of Oak Mere, Maer Pool, Quoisley Little Mere, Hatch Mere and Fodens Flash (from Savage, in preparation).

The associations of species of Corixidae with conductivity, organic matter (vegetation) and size–shape in lakes may be summarized thus:

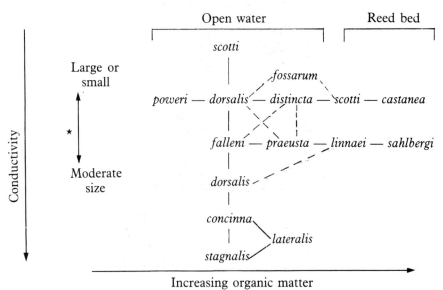

(*Applies only to *dorsalis* and *falleni*.)

Gerridae in North America and Finland are associated with a series of environmental components similar to those described in Corixidae. Spence (1983) analysed the vegetation structure of a habitat into three types, grass-sedge, floating open, and bulrush, each associated with the distribution of gerrids. He then identified five other factors related to species distribution, namely, invertebrate predator competition, habitat availability, surface conductivity, lake area and tree shelter. In Finland, *Limnoporus rufoscutellatus*, *Gerris lacustris*, *G. lateralis* and *G. odontogaster* form a succession associated with water vegetation, shore vegetation, depth and habitat stability (Vepsäläinen 1973). The developmental stages of Gerridae show differential habitat selection, with earlier instars living in denser vegetation than later instars and adults (Nummelin *et al.* 1984; Vepsäläinen & Nummelin 1986). Streams (1987) has described similar patterns of distribution in North American Notonectidae.

There is no doubt, in principle, of the occurrence of species specific patterns of ecological distribution among aquatic Hemiptera Heteroptera.

ECO-PHYSIOLOGY AND BEHAVIOUR

Respiration

Respiration may be by the uptake of dissolved oxygen from the surrounding water, as in the first few nymphal instars of Corixidae, but oxygen is normally taken direct from the atmosphere in both surface dwellers and subaquatic forms. The subaquatic species carry a bubble of air on their bodies which is renewed, in the majority, by periodic visits to the surface. This bubble may also act as a 'physical gill' by absorbing oxygen from the surrounding water. The physical gill alone suffices at low temperatures, for example under ice, in Corixidae and virtually permanently in *Ilyocoris cimicoides* (Popham 1960, 1964). The mode of respiration is clearly of ecological significance, but in most members of the taxon it appears to be a broadly-based adaptive strategy which has undergone little adaptive radiation in the different species. Thus our present knowledge of respiration appears to be of little help in the interpretation of ecological differences at species level. However, as indicated in Popham (1964) and as will be seen below, there are signs that detailed studies may reveal significant differences between species.

Salinity

Claus (1937) showed that *Sigara stagnalis* could survive only at salinities in the range $5-18\,\mathrm{g\,l^{-1}}$ (as NaCl equivalent), and thus delineated the fundamental niche in relation to water chemistry. Savage (1971c) showed that the realised niche (= actual ecological distribution) corresponded with this range in an inland saline lake in Cheshire. Two species of Corixidae inhabiting saline lakes in British Columbia, Canada, *Cenocorixa bifida hungerfordi* Lansbury and *C. expleta* (Uhler), also have differing fundamental niches in relation to salinity. They have partially overlapping realised niches, *C. bifida* 20–20,000 μS cm^{-1} and *C. expleta* 5,000–30,000 μS cm^{-1}. Osmoregulatory ability is responsible for the absence of *C. bifida* at high salinities but some other factor, probably parasitism, excludes *C. expleta* from low salinities (Scudder 1983). In general, Corixidae seem to be euryhaline (e.g. Savage 1971c).

Emergent vegetation

Quantitative estimates of Corixidae in three Swedish lakes showed that the three commonest species, *Arctocorisa germari*, *Glaenocorisa propinqua propinqua* and *Cymatia bonsdorffii*, have differing distributions in relation to increasing emergent vegetation. These differences were correlated with different behavioural characteristics. In laboratory experiments, *C. bonsdorffii* spent significantly longer times stationary ($p<0.01$) or in reeds ($p<0.01$) compared with swimming, whereas *G. p. propinqua* spent longer times swimming ($p<0.05$) or in open water ($p<0.01$). The number of visits to the water surface hour^{-1}, in four species were *A. germari* 0·52, *G. p. propinqua* 1·00, *Sigara distincta* 1·48, and *Hesperocorixa sahlbergi* 1·68 (Henrikson & Oscarson 1985). These four species are associated with increasing amounts of vegetation, respectively. It is probable that fish predation is related to these different patterns of behaviour, as will be seen in the relevant section below (p. 142).

Gregarious behaviour

Velia caprai is gregarious on the water surface but, in a Swedish population, spent approximately 70% of time on land. Females predominated in the aggregations on the water surface and activity in both sexes showed a significant positive correlation with diurnal temperature changes. It has been suggested that the gregarious behaviour may offer protection against fish predation (Brönmark et al. 1984, 1985; Erlandsson et al. 1988).

FEEDING AND DIET

The well developed piercing-sucking mouthparts, together with casual observations, have long indicated that all families of aquatic Hemiptera Heteroptera, except Corixidae, are predators which feed on living or recently dead animals. There have been conflicting opinions concerning the diet of Corixidae which will be considered later. The techniques of investigation have comprised direct observation in either field or laboratory and the examination of gut contents, visually or by serology and electrophoresis; all methods have limitations.

Surface dwellers

Walton (1943, 1981) observed the feeding behaviour and diet in the surface dwellers, *Hydrometra stagnorum* and Veliidae. Individuals are attracted to their prey by ripples produced at, or immediately below, the surface. They then pierce the prey, sometimes after a considerable struggle if it is large, and suck out the contents. *H. stagnorum* and *Microvelia* spp. feed mainly on water fleas (Cladocera), which may be lifted on the end of the rostrum, while *Velia* spp. have a more varied diet including spiders (Arachnidae), emerging midges (Chironomidae), stoneflies (Plecoptera) and mayflies (Ephemeroptera). In Sweden, *Velia caprai* probably fed mainly on small flies (Diptera) (Erlandsson *et al.* 1988).

Gerridae are generalised predators of invertebrates falling on the water surface and have been maintained, in laboratory experiments, on flies (Diptera) and moths (Lepidoptera) (Vepsäläinen & Nummelin 1986). Cannibalism may be important in limiting population size in *Gerris najas* when it occurs in oligotrophic waters where food is limited (Brinkhurst 1966) but it is unlikely that food is limiting in North American Gerridae in eutrophic waters (Spence 1986). The spatial isolation of instars in separate microhabitats, by interference of adults, larger nymphs and smaller nymphs, suggests that cannibalism is rare in Finnish populations of *Gerris argentatus*, *G. odontogaster*, *G. lacustris* and *Limnoporus rufoscutellatus* (Nummelin *et al.* 1984).

Notonectidae

The most thorough studies yet available are on Notonectidae. They were intended, primarily, as contributions to the general principles of predator-prey relationships but have yielded information and ideas which are directly relevant to the ecology of water bugs. Notonectidae detect their prey by visual and vibratory signals which, for instance, differ in *Notonecta glauca* and *N. maculata* (Walton 1943). The bugs wait either at the water surface or perched within the water body and, when prey are sufficiently close, pursue and capture them, by means of the raptorial fore and middle legs, prior to sucking out the contents (Giller 1986). *N. glauca* waits at different levels in the water in relation to temperature and oxygen concentration. At low temperature and high oxygen concentration more attacks were made on submerged prey and, as the distribution of prey was affected differentially by the physical conditions, any change affected the diet of the predators through the alteration of the spatial overlap (Cockrell 1984). In all species the rates of food extraction decreased exponentially with time and thus there was decreasing value in a prey item as a resource. The time spent feeding (handling time) was inversely proportional to prey density and hence a food resource is less completely used at high prey densities. However, the frequency of captures increases proportionally with prey density and thus individual predators consumed more food per unit time. There was also an increase in the frequency of captures through a particular foraging sequence and it has been suggested that a 'search image' is developed, based on recent memory (Cook & Cockrell 1978; Giller 1980). Thus, foraging behaviour is dependent on both innate and environmental parameters. Indeed, the differing patterns of foraging behaviour and food extraction in the different species are testimony to the importance of inherited adaptive strategies. Giller & McNeill (1981) have compared the adaptive strategies of *Notonecta glauca*, *N. obliqua* and *N. maculata* in relation to foraging behaviour and demonstrated convincingly that they are related to the types of habitat in which these species are found (Table 4).

Notonectidae are described as polyphagous carnivores (Cockrell 1984; Giller 1986). Giller (1986) investigated the feeding niches of third instar nymphs to adults of *N. glauca* and *N. marmorea viridis* from the same habitat, using an electrophoretic method of gut content analysis. Fingerprint esterase bands were identified for thirty-three potential prey types. The results showed that, of these, eighteen occurred in the natural diet, seven were taken under laboratory conditions, a further four were rejected under laboratory conditions while the remainder were not used in the laboratory or found in the natural diet. Examination of fig. 77 shows that the main diet (60%) of both species comprised *Cyclops* spp. and nymphs of *Cloeon dipterum* (L.), the former being restricted to earlier instars and replaced by the latter in the last two instars. However, there were significant differences in the use of *Sigara lateralis*, Ceratopogonidae and Dytiscidae. Thus, there is some evidence of resource partitioning between developmental stages of a species, and between species. There is also an

TABLE 4. INFORMATION ON FORAGING BEHAVIOUR AND HABITAT FACTORS IN
SPECIES OF NOTONECTIDAE

| | Species | | |
	N. glauca	N. obliqua	N. maculata
Foraging behaviour			
Position in water	Mid depth	Deep	Near surface
Attack frequency	Low	High	High
Handling time	Long	Short	Short
Extraction rate	Slow	High for small prey	High
Habitat factors			
Conductivity	High	Low	Variable
Vegetation	Dense	Moderate	Sparse
Prey density	High	Low	Low
Fish	Present	Absent	Absent

References: Giller & McNeill 1981; Savage unpublished observations.

indication that the diet may be more restricted than is implied by the term
'polyphagous'. However, from regression analysis there was a significant,
positive linear correlation between the numbers of individuals examined
(X) and the *total* number of prey types (Y) found in each instar,
$(Y = 1.446X - 4.516; r = 0.98, p < 0.001, n = 9)$, and the number of *different*
prey types (Z) slowly increased exponentially in relation to the numbers
examined (W), $(Z = 4.88 \ln W - 5.37; r = 0.87, p = 0.001, n = 9)$. The
number of different prey types found in an individual gut was usually one
in earlier instars and two in later instars; three prey types were very rare
(1%). Thus the polyphagous nature of the diet, or otherwise, will be
revealed only if a sufficiently large number of individuals is sampled.

Naucoridae

The foraging behaviour of *Ilycoris cimicoides* (Naucoridae) appears to be
similar to that found in Notonectidae. Adults and fifth instar nymphs were
provided with different densities of *Asellus aquaticus* L. and *Sigara (Corixa)*
striata. The numbers of prey killed and amounts of food ingested, estimated
as dry matter, increased significantly with prey density until a plateau was
reached. It also appeared likely that the bugs removed less dry matter
from individual prey items as prey density increased. More *A. aquaticus*
were caught than *S. striata*, a fact correlated with their individual value as
food resources (Venkatesan & Cloarec 1988). At present, it would be
inadvisable to attempt to deduce foraging behaviour under field conditions.

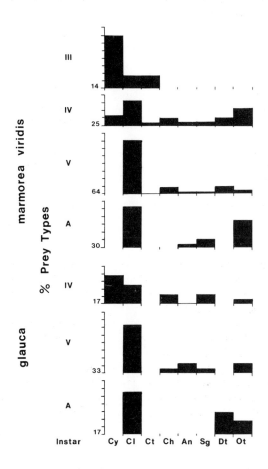

Fig. 77. The proportions of prey types found by electrophoresis of gut contents in the nymphal (III, IV or V) and adult (A) instars of *Notonecta glauca* and *N. marmorea viridis*. *Y* axis, scale divisions are 10% and the numbers of specimens examined are also given. *X* axis, prey types: *Cy, Cyclops* spp.; *Cl, Cloeon dipterum* larvae; *Ct,* Ceratopogonidae pupae; *Ch,* Chironomidae larvae; *An, Anopheles* spp. larvae; *Sg, Sigara lateralis*; *Dt* Dytiscidae larvae; *Ot,* others or unknown. (Data from Giller 1986).

Corixidae

The Corixidae lack a distinct rostrum but they possess stylets and are capable of piercing and sucking a food source. In addition, they possess buccopharyngeal teeth, varying somewhat in different species, which appear to be used for masticating solid food gathered with the palae (Elliott & Elliott 1967; Sutton 1951). Thus, it would appear that the feeding habits and diets of Corixidae are likely to be different from other water bugs. They have been described as omnivores, carnivores, detritivores and algal feeders by a variety of authors (reviews by Bakonyi 1978; Jansson & Scudder 1972; Popham *et al.* 1984; Reynolds 1975). These differing views are not necessarily mutually exclusive but rather they represent attempts at generalisations from limited evidence. It is now becoming clear that there are considerable differences in diet between species of Corixidae.

Early detailed observations of British Corixidae indicated that *Cymatia bonsdorffii* and *Glaenocorisa propinqua* were carnivores (Walton 1943). Later, it was shown that *Corixa punctata*, *C. panzeri*, *Sigara dorsalis* and *S. falleni* were omnivores, under laboratory conditions, feeding on invertebrates, algae and detritus. Moreover, solid food was ingested, in addition to the usual technique of piercing and sucking (Sutton 1951). Popham *et al.* (1984) attempted to define the natural diets of twenty-one species by visual examination of the gut contents of 3502 specimens, since the occurrence of small pieces of plants and animals showed that food was partly ingested. Corixids pierce and suck when feeding on prey but they also ingest solid food which is masticated by the buccopharyngeal teeth; some small organisms may be ingested whole. It was realised that only a general classification would be possible and therefore each individual was assigned to one of four categories: animal, algal, detritus, or mixed feeder. It should be appreciated that, for example, small quantities of algae might be present in a specimen that was predominantly an animal feeder and classified as such (fig. 78). *Cymatia bonsdorffii*, *Corixa panzeri* and *C. dentipes* were carnivores while *C. punctata* was a carnivore that was sometimes an omnivore. Most of the other species listed were omnivores but *Callicorixa wollastoni*, *Hesperocorixa linnaei* and *H. sahlbergi* were sometimes detritus feeders. All specimens of *Sigara lateralis* and *S. stagnalis* were detritivores. It is self evident that there were wide variations between the diets of individuals of the same species. Indeed, there was a distinct difference between the sexes in *S. falleni*.

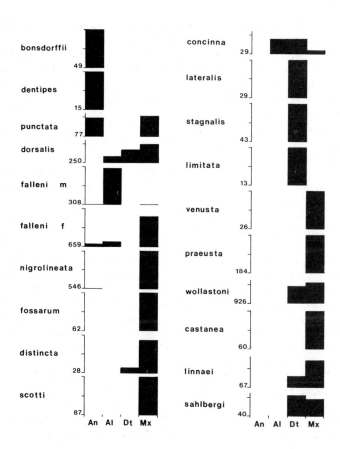

Fig. 78. The proportions of specimens of species of Corixidae containing four categories of food: *An*, animal; *Al*, algal; *Dt*, detritus; *Mx*, mixed. *Y* axis, scale divisions are 50% and the numbers of specimens examined are also given; results on males (*m*) and females (*f*) are shown separately for *S. falleni*. (Data from Popham *et al.* 1984).

The diets of Hungarian specimens of *S. lateralis* varied with time during the summer (Bakonyi 1978) while the diets of two Canadian species differed according to whether they were allopatric or sympatric (Reynolds & Scudder 1987a, 1987b). Cannibalism and interspecific predation have been noted and may be important factors in limiting population size under crowded conditions (Pajunen & Ukkonen 1987).

There is still much uncertainty concerning the feeding habits of British Corixidae but recognizable patterns may be established if they are considered in relation to the functional morphology of the head and fore legs, together with their ecological distribution. *Glaenocorisa propinqua* is a denizen of deep open water, feeding on pelagic invertebrates for which it trawls with its palae. *Cymatia bonsdorffii* perches on vegetation, rather like Notonectidae, from which it pursues its prey, using the cylindrical fore tarsi and middle legs to capture it (figs 38, 57; Henrikson & Oscarson 1978b; Walton 1943). These two species have particularly well developed eyes and, together with *Arctocorisa carinata* and *A. germari*, have long fore coxae. The majority of Corixidae are benthic feeders and collect their food at or near the substratum. There are progressive changes in the position of the head and the shape of the fore legs, associated with omnivores and detritivores. In contrast, *Sigara dorsalis*, *S. distincta* and *S. falleni*, which tend towards algal feeding, show a trend towards larger, flatter palae (fig. 50; Popham *et al.* 1984). The omnivores and algal feeders tend to be associated with habitats containing little organic matter in their substrata while the detritus feeders are associated with high contentrations of organic matter (figs 75, 76; Macan 1938, 1954a, 1967, 1970; Savage 1971c, 1982b; Savage & Pratt 1976).

PREDATION

The most important predators of aquatic Hemiptera Heteroptera are almost certainly other water bugs and fish, although *Gammarus tigrinus* Sexton is important in the few habitats where it occurs (Henrikson & Oscarson 1978b, 1981; Macan 1965b, 1965c, 1976; Oscarson 1987; Pajunen & Ukkonen 1987; Popham 1966; Savage 1981, 1982a).

Popham (1966), in a most thoughtfully designed field experiment, introduced 1000 *Sigara nigrolineata* into a pond which contained the three-spined stickleback (*Gasterosteus aculeatus* L.). *S. nigrolineata* is very variable in pigmentation and it was possible to divide the specimens into three groups according to colour intensity, viz: light, medium or dark. The three colour types were introduced in equal numbers, but after seven days there were significantly relatively greater numbers of the dark type, and the whole population had been reduced by 77%. Sticklebacks were seen attacking and eating corixids. They attacked moving specimens of all colours and light-coloured stationary specimens on the dark substratum of the pond. However, motionless dark specimens were not attacked. Thus, both colour and behaviour were involved in differential survival although the overall mortality was very high.

Gammarus tigrinus was seen attacking and eating corixid nymphs in a slightly saline lake (Savage 1981, 1982a). Laboratory experiments confirmed that first, second and third instar nymphs of *Sigara dorsalis*, *S. falleni* and *S. lateralis* were particularly vulnerable while later nymphal instars and adults were rarely attacked successfully. Dishes containing open water or with artificial vegetation were used; mortality was much lower in the latter.

Arctocorisa carinata and *Callicorixa producta* are sympatric in rock pools in Finland. Both species show interspecific and intraspecific (cannibalistic) predation. In laboratory experiments the killed nymphs showed a number of ragged holes on the dorsal surface and the body contents had gone. In each species intraspecific predation showed a positive linear correlation with prey density but predator efficiency remained constant. There was wide variation between individuals in their efficiency as predators but *A. carinata* was significantly more efficient than *C. producta*. Small nymphs of *A. carinata* were also more adept in evading larger nymphs of both species. It was clear that *A. carinata* was the more efficient predator of the two and hence one might predict that *C. producta* would become extinct in these habitats. However, the heterogeneous, discontinuous and temporary nature of rock pools allowed *C. producta* to persist as a fugitive migrant species (Pajunen 1979a, 1979b, 1982; Pajunen & Ukkonen 1987). *C. producta*, a ready migrant, thus possessed a compensatory adaptive strategy unrelated to predator efficiency.

There are associations between the presence of fish and the species composition and spatial distribution of water bugs which may be attributed to predation. For example, Macan (1976) made an experimental study of trout predation on water bugs in Hodson's Tarn, a small artificial moorland fishpond in the English Lake District. The tarn's population of brown trout (*Salmo trutta* L.) was removed between 1948 and 1955, when observations on aquatic plants and macroinvertebrates began. For the twenty-one years of continuous study, conditions in Hodson's Tarn are divisible into four distinct periods: (1), 1955-1960 when fish were absent; (2), 1961-1965 when fish were present (500 trout were introduced each autumn in 1960 and 1961); (3), 1966-1967 and (4), 1968-1975 when fish were again absent. In periods (1) to (3), aquatic vegetation covered most of the tarn's area; in period (4) the centre of the tarn was clear from vegetation. Nineteen species of water bugs were collected during the twenty-one years but only five were found in large numbers. *Notonecta obliqua*, *Sigara scotti* and *Hesperocorixa castanea* were common during period (1) when fish were absent (fig. 79, a1). When fish were introduced (period 2) their distribution changed in relation to the vegetation, and *N. obliqua* virtually ceased to be a breeding species (fig. 79, a2). Immediately after the fish had been removed again (period 3) the community returned to that of the first period (fig. 79, a1). When the central vegetation atrophied (period 4), *Sigara distincta* and *Cymatia bonsdorffii* colonised that part of the tarn (fig. 79, a3). Examination of trout stomachs showed that bugs had been eaten (Macan 1965b, 1965c, 1976).

A similar study of a community comprising *Glaenocorisa propinqua propinqua*, *Sigara distincta* and *S. scotti* in a Swedish lake showed a marked change in species composition and distribution after the introduction of roach (*Rutilus rutilus* (L.)) (fig. 79, b1, b2) (Oscarson 1987). Likewise, Oak Mere, Cheshire, which has a large population of perch (*Perca fluviatilis* L.), yielded very few corixids when the water level did not reach the marginal vegetation (Savage & Pratt 1976) but large numbers of *Sigara dorsalis* were found when the level rose and the marginal vegetation was flooded (Savage 1982b, unpublished observations).

The acidification of many Scandinavian lakes, owing partly to atmospheric pollution, has offered further opportunities for studying the effects of fish predation on Corixidae. *Glaenocorisa propinqua propinqua* was rare in Swedish lakes until the 1970s but during that decade it greatly increased its range in acidified lakes where fish populations had become depleted or extinct. Experimental studies based on the introduction of fish and the examination of fish stomachs showed that perch (*Perca fluviatilis*) and roach (*Rutilus rutilus*) fed on *G. p. propinqua* and probably controlled its

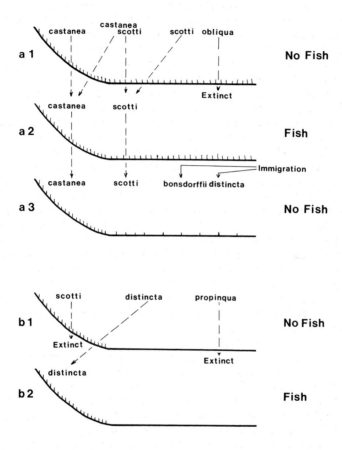

Fig. 79. A schematic representation of the effects of vegetation (aquatic macrophytes) and fish predation on the occurrence and spatial distribution, with time, of some species of Notonectidae and Corixidae in *a1, a2, a3*, Hodson's Tarn, England; *b1, b2*, Lake Lilla Stockelidsvatten, Sweden. The short vertical lines indicate distribution and density, according to closeness, of vegetation. (Data from Macan 1965b, 1965c, 1976; Oscarson 1987).

occurrence. These changes have resulted in Corixidae becoming the top aquatic predators in acidified lakes, taking the role of fish, and, in turn, have themselves become a new and important source of food for the Goldeneye (*Bucephala clangula* (L.)) and other species of water fowl (Henrikson & Oscarson 1978b, 1981).

There can be no doubt of the significance of the relationships between the occurrence and distribution of Corixidae and such factors as water body morphometry, chemical composition, distribution of vegetation and the nature of the substratum (e.g. Macan 1938, 1954a; Savage 1982b). However, there is now equally convincing evidence that fish predation of Corixidae, and perhaps other aquatic bugs, by perch (*Perca fluviatilis*), roach (*Rutilus rutilus*), brown trout (*Salmo trutta*) and sticklebacks (*Gasterosteus aculeatus*) plays an important role. Indeed, it seems highly probable that fish predation is a key factor in the maintenance of the so-called 'normal' distribution of Corixidae within a habitat. However, the importance of factors such as water chemistry should not be forgotten. The four commonest species in twenty-six acidified lakes in Norway and Sweden were *Glaenocorisa p. propinqua* (21% of records; 73% of lakes), *Sigara distincta* (19%, 65%), *Cymatia bonsdorffii* (15%; 54%) and *S. scotti* (12%; 42%) (Henrikson & Oscarson 1981). In Hodson's Tarn, a base-poor fishpond in England, the four commonest species were *Hesperocorixa castanea* (48% of total individuals collected), *S. scotti* (32%), *S. distincta* (14%) and *C. bonsdorffii* (2·4%). There is a marked similarity between the corixid communities of these two geographically separated localities, with similar chemical compositions and no fish.

It has sometimes been suggested that water bugs have chemical defences against potential predators, produced from superficial secretory glands. The following British species produced the substances indicated: *Notonecta glauca* and *Ilyocoris cimicoides*, p–hydroxy benzaldehyde and p–hydroxy benzoic acid methyl ester; *Sigara falleni* and *Corixa dentipes*, trans–4–oxo–hex–2–enal; *Plea leachi*, hydrogen peroxide. These substances may be secreted when an individual is attacked by a fish and there is some evidence that it may be rejected (Aldrich 1988; Kerfoot & Sih 1987). Present ecological evidence suggests that these substances give little protection against fish predation, as described above.

PARASITISM

Casual observations suggest that parasitism may be an important factor in the ecology of British water bugs. I have frequently found communities of Corixidae which were heavily infested with mites (Hydrachnidae) and a population of *Arctocorisa germari* in Barbrook Reservoir was infested with an endoparasite which was probably a gregarine protozoan (Crisp 1962b). The hymenopteran *Tiphlodytes gerriphagus* (Marchal) is parasitic on the eggs of a number of species of North American Gerridae. It was found that the parasite population grew less well in unstable habitats and this may explain the restriction of *Limnoporus dissortis* (Drake & Harris) to such places (Spence 1986). *Hydrachna conjecta* Koenike is a parasite of many European species of Corixidae. Its biology was studied in populations of *Cymatia coleoptrata*, *Sigara striata* and *S. falleni* in the Netherlands. Adults and nymphs of *H. conjecta* fed directly upon *Sigara* eggs whereas larvae were parasitic on adult Corixidae. The larvae attach themselves to the corium on the ventral surface of the hemielytra of these three species, and pierce the haemocoel. However, they fail to develop in *S. falleni* which is described as 'immune'. A fully grown larva forms 13% of the total volume of *S. striata* and 31% of *C. coleoptrata* (Davids 1973). *H. conjecta* consumes about 200 eggs ($4 \cdot 5$ day^{-1}) of *S. striata* during its lifetime and, with 10% parasitism, reduces egg production in the spring generation by 25–30% (Davids *et al.* 1978). Observations on *S. scotti* also suggest that female fecundity may be reduced (Crisp 1959).

FLIGHT POLYMORPHISM, OVOGENESIS AND MIGRATION

Wing polymorphism, wing muscle development polymorphism, variations in flight threshold and possibly a whole range of associated physiological and behavioural factors, including migration, are normal in aquatic Hemiptera Heteroptera.

Flightlessness is a general adaptation to the freshwater environment. However, freshwater habitats are basically discontinuous and vary, through an intergrading series, from large stable permanent habitats to small unstable temporary habitats. For species in permanent habitats, a tendency to disperse is disadvantageous for survival since they are unlikely to find an equally suitable place, whereas it is a prerequisite for survival in temporary habitats. Some habitats are stable for one generation in a life cycle but not for another. Under such conditions winged and wingless forms may each offer better chances of survival in different seasons. The environment, in general, has dynamic properties and, thus, almost all species are capable of producing both winged and wingless forms. The proportions and temporal occurrence of these forms are related to environmental factors.

The wing lengths of British adult Gerridae and Veliidae may be divided into at least six different categories from wingless (apters) through micropters, sub-brachypters, brachypters, sub-macropters to fully winged (macropters) (Brinkhurst 1959a). However, the major function of wings is to provide a means of migratory dispersal and details of wing length are less important than the ability and tendency to fly. Examination of a population of *Gerris odontogaster* in the English Lake District showed that 39% ($n = 118$) of macropters lacked wing muscles and were unable to fly (Brinkhurst 1959a). Clearly, dispersal ability cannot be deduced directly from external morphological features alone. A study of four Canadian species of Gerridae has demonstrated a significant negative correlation between flight threshold, degree of degeneration of wing muscles and the *proportion* of macropters in a population. In other words, there is an increasingly lower dispersal ability as the proportion of macropters decreases (Fairbairn 1986; Fairbairn & Desranleau 1987). It is probable that similar associations occur in European species (cf. Brinkhurst 1959a, 1963; Vepsäläinen 1974). This evidence supports the proposition of Young (1965a, 1965b), for Corixidae, and Andersen (1973), for Gerridae, that the dispersal or migratory ability of a species depends not only on the proportion of macropters but also, and perhaps primarily, on the dispersal capability of the macropterous morph.

Seasonal morphs and dispersal

It has already been shown that water bugs have one or two generations per year i.e. are univoltine or bivoltine (fig. 67). A study of the relationship of the dispersal ability of a species with the nature of its habitat is inseparable from a study of voltinism in that species, since flight polymorphism may vary during the annual life cycle. By extrapolation from the work described above, it is possible to estimate dispersal ability, as a function of wing polymorphism, in species occurring in the British Isles; this has been attempted (Table 5). Clearly, there are negative associations between high dispersal ability and habitat stability. There is also a distinct tendency, in the bivoltine species, towards the production of a sedentary reproductive summer generation followed by a winter population with high dispersal ability. The details of this relationship appear to be modified according to the degree of habitat stability.

TABLE 5. THE DISPERSAL ABILITY OF SPECIES OF VELIIDAE AND GERRIDAE IN RELATION TO THE STABILITY OF THEIR BREEDING HABITATS

Species	Breeding habitat	Dispersal ability	
Univoltine			
Velia caprai			
Velia saulii	Stable	Low	
Gerris najas			
Limnoporus rufoscutellatus†	Unstable–stable	High	
Gerris costai poissoni	Unstable	High	
Gerris gibbifer			
Bivoltine		Summer generation	Winter generation
Gerris paludum	Moderately stable	Low	High
Gerris lacustris	Unstable–stable	Moderate	Fairly high
*Gerris lateralis asper**			
Gerris odontogaster	Unstable	Moderate	High
Gerris argentatus			
Gerris thoracicus			

† *L. rufoscutellatus* based on Finnish data only;
* *G. lateralis asper* may be univoltine and polymorphic also.
References: Brinkhurst 1959a, 1963; Vepsäläinen 1974.

Flight polymorphism has been thoroughly studied in British Corixidae (Young 1965a, 1965b). Thirty-three species were investigated, of which four were polymorphic in both wing and wing muscle development, twenty-two were polymorphic in wing muscle development only, and all specimens of seven species had fully developed wing and wing muscles. The majority of species have four distinct morphs: early flightless and normal forms characterised by light pigmentation appear in early summer, followed by the main flightless and normal forms with darker pigmentation in late summer and autumn. As in Gerridae, the proportions of these morphs vary between species and, during the summer part of the cycle, within some species. The proportions are similarly associated with habitat stability. The differences may be exemplified by a comparison of *Sigara dorsalis*, a species often associated with large permanent habitats, and *S. scotti*, a species more associated with small, less stable habitats (fig. 80). There is a much higher proportion of the main normal flying morph in *S. scotti*, but the proportions are similar in the two species of the early flightless morphs during the reproductive season.

As already stated, Macan (1954a) grouped species of Corixidae into a series of habitat types according to a variety of environmental factors. Young (1965b) paid particular attention to three pairs of species from three types of habitat. *Sigara dorsalis* and *S. falleni* from the lake–river group, *S. scotti* and *Hesperocorixa castanea* from the acid unproductive pond group, and *S. lateralis* and *S. nigrolineata* from productive ponds. Flightless morphs were often abundant in the first pair of species while normal morphs predominated in the remaining two pairs.

Wing length is controlled by photoperiodic induction in *Gerris odontogaster* (Vepsäläinen 1971a, 1971b). A particular Finnish population was apparently bivoltine (but see later) and dimorphic. The first generation comprised both micropters and macropters while the second generation comprised virtually macropters only (0·5% micropters). Laboratory and field experiments showed that there was wide intraspecific variation in rates of development (cf. fig. 67) and that only those individuals which were exposed to long, lengthening days, up to a maximum of 18 hours and 22 minutes during the first four nymphal instars, developed into micropters. In contrast, individuals exposed to shortening days, 5 minutes day^{-1} from an initial long day of 22 hours and 40 minutes, developed into macropters. Thus, change of day length was more important than the absolute day length. This response ensures that the mechanism will function at a variety of latitudes except in the extreme north, about latitude 64°N, where individuals do not develop beyond the fourth instar until after the summer solstice, 22 June; then only macropters are produced (Ekblom 1950).

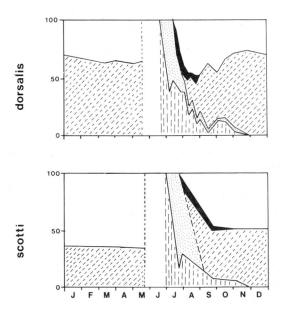

Fig. 80. The proportions of morphs during the annual cycle of *Sigara dorsalis* from a stable habitat (above) and *S. scotti* from an unstable habitat (below). Very few adults were caught during late May and June and hence this period is omitted. Key: vertical shading, newly moulted adults; dotted, early flightless morph; diagonal, main flightless morph; solid, developing normal morph; unshaded, normal morph. (Data from Young 1965b).

In Corixidae there is evidence that temperature may play a part in the environmental control of flight polymorphism. *Sigara scotti* was found in a pond with a temperature of *c.* 24°C and also in adjacent small isolated pools at *c.* 31°C; the former were dimorphic while the latter were monomorphic flying forms. Flightless morphs of *S. lateralis* developed wing muscles when exposed to high temperature and/or pollution (Young 1965b). However, it must be emphasised that these statements are based on deductions from field data and were not the results of rigorous laboratory experiments as with *G. odontogaster*.

Ovogenesis

Ovarian development is so closely linked with flight polymorphism in Corixidae and Gerridae that it must be considered in this section. The initiation of ovarian development in overwintered adults varies widely from February to May, roughly in relation to latitude and the species concerned. It is frequently accompanied by flight muscle degeneration in the normal overwintered morphs. In contrast, the summer cessation of ovarian development, diapause, occurs immediately following the summer solstice, 22 June, after which the days shorten. Diapause occurred in both normal and flightless morphs in Corixidae but only in macropters in *Gerris odonotogaster*. In *G. odontogaster* there is a common gene pool in overwintering macropters. These give rise, in the following year, to the first generation micropters showing ovarian development, and to diapausing macropters. The second generation consists of diapausing macropters only. Micropters do not overwinter but the macropters of both generations provide a common overwintering population capable of dispersal. It seems almost certain that ovarian development and wing polymorphism are controlled by photoperiodic induction in Gerridae and Corixidae (Brinkhurst 1959a; Pajunen 1970; Vepsäläinen 1971a, 1971b; Young 1965b). The present evidence indicates that all Corixidae and Gerridae have either a univoltine or partially bivoltine life cycle. Complete bivoltinism has yet to be demonstrated. It is clear that the deductions of earlier workers, based on casual field observations, should be treated with extreme caution.

Migration

Brown (1951b) demonstrated that the frequency of occurrence of immigrant Corixidae in a new pond was related to the type of habitat in which they normally lived rather than their relative abundance in the surrounding area. Regression analysis of his data shows that there was a significant positive correlation between the percentage of temporary habitats (X) in which given species were found and the natural logarithm of their relative migration rates $(Y,$ the number of immigrants recorded in the new pond, divided by the number in samples collected from the surrounding area), viz: $\ln Y = a + bX$; $r = 0.76$, $p < 0.02$, $n = 8$. Thus, the migration rate of species increases exponentially in relation to habitat instability.

The main periods of migration of Corixidae in Central Lancashire (England) (fig. 72, NR) are April-May and late July-September, periods associated with anticyclonic weather conditions. Under such conditions temperatures in the littoral zone of lakes and ponds rise, reaching a maximum in the afternoon. Corixidae follow the increasing temperature gradient and accumulate near the shore. They normally visit the water surface, but do so with exponentially increasing frequency as temperature rises. Emigration occurs increasingly under these circumstances, in individuals that can fly. Once airborne, they are attracted by highly reflective surfaces (Popham 1964). The previous account of developmental processes suggests that spring and autumn are periods when flight ability in a population is likely to be highest, especially in species of temporary habitats.

Natural selection of morphs

Flight polymorphism, ovarian development and presumably an associated range of physiological, behavioural and life cycle traits are correlated with environmental factors. These responses must depend upon the inherited potentiality of the individuals comprising populations and species. It appears that multigenic control mechanisms provide the most probable explanation although control of particular aspects by a single pair of alleles has been suggested by a number of the earlier workers (Zera 1985). There is evidence that the different morphs have different degrees of inherent fitness. In *Gerris lacustris* macropters exhibited higher overwintering survivorship than micropters (Vepsäläinen 1974) while flightless morphs of *Sigara dorsalis, S. falleni* and *S. scotti* survived longer and laid more eggs than normal morphs when kept under adverse conditions in laboratory experiments (Young 1965b). In contrast, apterous and macropterous morphs of the North American species, *Gerris remigis* Say, showed no significant differences in development times, the proportion breeding without diapause and overwintering survivorship. However, pre-diapause macropterous females had a significantly shorter pre-oviposition period than apterous females whereas post-diapause macropters began reproducing later than apters. These results suggest that macropters may be at a selective advantage in warm habitats which favour pre-diapause reproduction, but that apters should be favoured in the preferred cool, lotic habitats (Fairbairn 1988).

Thus, wing polymorphism is a morphological 'marker' of a whole suite of related adaptive strategies which reflect both a general tendency towards monomorphic apterism in stable habitats and the balanced polymorphism of less stable habitats.

COMMUNITY STUDIES

There have been a number of long term studies of water bug communities. Pajunen (e.g. 1970, 1977, 1979a, 1979b, 1981, 1982, 1983) has revealed the complexity of interrelationships between *Arctocorisa germari* and *Callicorixa producta* in the unstable rock pools of the coast of Finland. In North America, Spence (e.g. 1981, 1983, 1986) has studied Gerridae and Scudder (e.g. 1976, 1983) Corixidae in saline habitats. Two such studies have been made in the British Isles, one in Hodson's Tarn in the English Lake District (e.g. Macan 1965b, 1965c, 1976) and the other in the slightly saline Watch Lane Flash, Cheshire (e.g. Savage 1971c, 1985). The former has been quoted in relation to predation and hence the latter is used as a basis for a discussion of possible causal relationships affecting population dynamics and distribution of species.

In Watch Lane Flash, between 1967 and 1979 the commonest macroinvertebrates were Corixidae and Gammaridae. The numbers collected in each year are shown together with certain environmental factors (fig. 81). From 1967 to 1970 the community comprised large numbers of *Gammarus duebeni* Liljeborg and *Sigara concinna*. *S. dorsalis* and *S. stagnalis* were present in small numbers together with *Nepa cinerea* and a very few others. This community was associated with a dense sward of *Myriophyllum spicatum* L., covering an area of four hectares in water averaging 0·3 m deep and summer conductivities of 7,000–8,000 μS cm^{-1}k25 (Savage 1971c, 1979a). In 1971-1972 the numbers of all bug species decreased, except for *S. stagnalis* in 1971, followed by an increase in numbers of *S. lateralis* which was first recorded in 1970. This change was associated with an anthropogenic modification of the saline source, a rise in summer conductivities to 9,000–10,000 μS cm^{-1} k 25 and the atrophy of *M. spicatum* which left a mass of decaying plants. *S. concinna* laid eggs exclusively on plants while *S. stagnalis* and *S. lateralis* laid eggs somewhat indiscriminately (Savage 1979b). It seems probable that the rise in salinity was caused by human interference and in turn caused the death of *M. spicatum*, which removed the oviposition sites of *S. concinna*. *S. stagnalis* was abundant in a nearby highly saline lake and was seen emigrating. Until 1974, summer conductivities remained roughly constant at 9,000 μS cm^{-1}k25, 5 g l^{-1} NaCl, and the numbers of *S. lateralis* increased markedly while *S. stagnalis* increased marginally. The salinity was at the lower limit for *S. stagnalis* (Claus 1937) which may have impaired its competitive ability in relation to *S. lateralis* owing to osmoregulatory stress. On the other hand, *S. lateralis* is associated with fouling by cattle and decaying plants may provide a similar food resource (Macan 1954a; Popham *et al.* 1984;

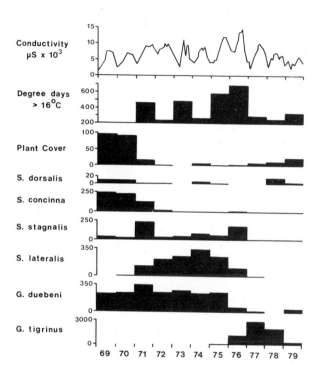

Fig. 81. The mean monthly conductivity ($\mu S\,cm^{-1} \times 10^3\,k\,25$), degree days year^{-1} above 16 °C, % cover year^{-1} of higher plant species, and numbers 100 standard net sweeps^{-1} of four corixid species ($S = Sigara$) and two species of *Gammarus* in an inland saline lake, Watch Lane Flash. (Adapted from Savage 1985).

Savage & Pratt 1976). In 1975 the accidental introduction of *Gammarus tigrinus* coincided with sparse vegetation, high summer temperatures and high conductivities in 1975-1976 which provided optimum conditions for its growth. It proved to be a voracious predator of *G. duebeni* and corixid nymphs and their numbers fell to a very low level (Savage 1981, 1982a). Salinity was significantly correlated with net rainfall and it seems likely that climatic factors initiated this causal chain (Savage 1977). Net rainfall was higher and summer temperatures lower in subsequent years. Conductivity decreased, the vegetation began to recover, the numbers of *G. tigrinus* fell and there were signs of an increase in numbers of Corixidae (fig. 81). Annual collections were continued; by 1984 *M. spicatum* was common again and the corixid community was represented principally by *S. concinna* and *S. lateralis*. Environmental factors and the macroinvertebrate community remained constant until 1988.

Thus, there are indications that many factors were responsible for the population fluctuations noted in Watch Lane Flash. The general interrelationships between patterns of distribution and causal processes have been eloquently propounded by Spence (1983). He pointed out that similarities of pattern do not imply similarity of processes. Indeed, although many ecological patterns are well established, the investigation of causal processes is only just beginning and is the major immediate task.

ACKNOWLEDGEMENTS

My grateful thanks are accorded to all of the many colleagues who so generously gave their time and effort in dealing with my enquiries. I hope that those not specifically mentioned below will forgive the omission as it does not imply any failure of appreciation. Dr J. M. Elliott, who invited me to undertake this task, has dealt patiently with my many requests. T. I. Furnass has photographed the figures and Dr D. W. Sutcliffe has edited and prepared the key for publication. Dr I. Lansbury, of Oxford University Museum, has been a constant source of assistance and good advice. Dr I. D. Wallace, of Liverpool Museum, and E. G. Philp, of Maidstone Museum, have made comments and provided large numbers of specimens. Dr J. H. Crothers, editor of *Field Studies*, has kindly allowed me to use parts of an article accepted for that journal. Finally, a special word of thanks to J. Cartwright and my wife, Pauline, for doing so much to smooth the path.

REFERENCES

Aiken, R. B. (1982). Sound production and mating in a waterboatman *Palmacorixa nana* (Heteroptera: Corixidae). *Animal Behaviour*, 30, 54–61.
Aiken, R. B. (1985). Sound production by aquatic insects. *Biological Reviews*, 60, 163–211.
Aldrich, J. R. (1988). Chemical ecology of the Heteroptera. *Annual Review of Entomology*, 33, 211–38
Allen, K. R. (1951). The Horokiwi Stream: a study of a trout population. *Fisheries Bulletin, New Zealand Marine Department*, 10, 1–231.
Andersen, N. M. (1973). Seasonal polymorphism and developmental changes in organs of flight and reproduction in bivoltine pond skaters (Hemiptera, Gerridae). *Entomologica Scandinavica*, 4, 1–20.
Andersen, N. M. (1982). *The semiaquatic bugs (Hemiptera, Gerromorpha)*. Ento-monograph 3, Scandinavian Science Press, Klampenborg, Denmark.
Arnqvist, G. (1988). Mate guarding and sperm displacement in the water strider *Gerris lateralis* Schumm. (Heteroptera: Gerridae). *Freshwater Biology*, 19, 269–74.
Bakonyi, G. (1978). Contribution to the knowledge of the feeding habit of some water boatmen: *Sigara* spp. (Heteroptera: Corixidae). *Folia Entomologica Hungarica*, 31, 19–24.
Behr, H. (1988). Eine weitere Falle zur quantitativen Erfassung luftatmender Wasserinsecten. *Archiv für Hydrobiologie*, 112, 631–8.
Brinkhurst, R. O. (1959a). Alary polymorphism in the Gerroidea (Hemiptera-Heteroptera). *Journal of Animal Ecology*, 28, 211–30.
Brinkhurst, R. O. (1959b). The habitats and distribution of British *Gerris* and *Velia* species. *Journal of the Society for British Entomology*, 6, 37–44.
Brinkhurst, R. O. (1959c). A description of the nymphs of British *Gerris* species (Hemiptera-Heteroptera). *Proceedings of the Royal Entomological Society of London* (A), 34, 130–6.
Brinkhurst, R. O. (1963). Observations on wing-polymorphism in the Heteroptera. *Proceedings of the Royal Entomological Society of London* (A), 38, 15–22.
Brinkhurst, R. O. (1966). Population dynamics of the large pond-skater *Gerris najas* Degeer (Hemiptera-Heteroptera). *Journal of Animal Ecology*, 35, 13–25.
Brönmark, C., Malmqvist, B. & Otto, C. (1984). Anti predator adaptations in a neustonic insect (*Velia caprai*). *Oecologia*, 61, 189-191.
Brönmark, C., Malmqvist, B. & Otto, C. (1985). Dynamics and structure of a *Velia caprai* (Heteroptera) population in a South Swedish Stream. *Holarctic Ecology*, 8, 253–8.

Bröring, U. & Niedringhaus, R. (1988). Zur Ökologie aquatischer Heteropteren (Hemiptera: Nepomorpha) in Kleingewässern der ostfriesischen Insel Norderney. *Archiv für Hydrobiologie,* 111, 559–74.

Brown, E. S. (1943). A contribution towards an ecological survey of the aquatic and semi-aquatic Hemiptera-Heteroptera (water bugs) of the British Isles; Anglesey, Caernarvon and Merioneth. *Transactions of the Society for British Entomology,* 8, 169–230.

Brown, E. S. (1948). A contribution towards an ecological survey of the aquatic and semi-aquatic Hemiptera-Heteroptera (water bugs) of the British Isles; dealing chiefly with the Scottish Highlands, and East and South England. *Transactions of the Society for British Entomology,* 9, 151–95.

Brown, E. S. (1951a). The identity of British *Velia* (Hem. Veliidae), with an account of a species new to Britain. *Entomologist's Monthly Magazine,* 87, 297–306.

Brown, E. S. (1951b). The relation between migration rate and type of habitat in aquatic insects, with special reference to certain species of Corixidae. *Proceedings of the Zoological Society of London,* 121, 539–45.

Carrick, T. R. & Sutcliffe, D. W. (1982). Concentrations of major ions in lakes and tarns of the English Lake District (1953–1978). *Occasional Publications of the Freshwater Biological Association* No. 16, 170 pp.

Chandler, T. J. & Gregory, S. (1976). *The climate of the British Isles.* Longman, London.

Claus, A. (1937). Vergleichend-physiologische Untersuchungen zur Ökologie der Wasser-wanzen. *Zoologische Jahrbücher,* 58, 365–432.

Cobben, R. H. & Pillot, H. M. (1960). The larvae of Corixidae and an attempt to key the last larval instar of the Dutch species (Hem., Heteroptera). *Hydrobiologia,* 16, 323–56.

Cockrell, B. J. (1984). Effects of temperature and oxygenation on predator-prey overlap and prey choice of *Notonecta glauca. Journal of Animal Ecology,* 53, 519–32.

Cook, R. M. & Cockrell, B. J. (1978). Predator ingestion rate and its bearing on feeding time and the theory of optimal diets. *Journal of Animal Ecology,* 47, 529–47.

Crisp, D. T. (1959). Hydracarines and Nematodes parasitizing *Corixa scotti* (D. and S.) (Hemiptera) in western Ireland. *Irish Naturalists' Journal,* 13, 88–92.

Crisp, D. T. (1962a). Estimates of the annual production of *Corixa germari* (Fieb.) in an upland reservoir. *Archiv für Hydrobiologie,* 58, 210–33.

Crisp, D. T. (1962b). Observations on the biology of *Corixa germari* (Fieb.) (Hemiptera Heteroptera) in an upland reservoir. *Archiv für Hydrobiologie,* 58, 261–80.

Davids, C. (1973). The water mite *Hydrarachna conjecta* Koenike, 1895 (Acari, Hydrachnellae), bionomics and relations to species of Corixidae (Hemiptera). *Netherlands Journal of Zoology,* 23, 363–429.

Davids, C., Al, M. E. & Blaauw, J. (1978). Influence of the water mite *Hydrarachna conjecta* on the population of the corixid *Sigara striata* (Hemiptera). *Verhandlungen der Internationalen Vereinigung für theoretische und angewandte Limnologie*, **20**, 2613–16.

Drake, C. J. (1920). An undescribed water-strider from the Adirondacks. *Bulletin of the Brooklyn Entomological Society*, **15**, 19–21.

Ekblom, T. (1950). Über den Flügelpolymorphismus bei *Gerris odontogaster* Zett. *Notulae Entomologicae*, **30**, 41–9.

Elliott, J. M. (1977). Some methods for the statistical analysis of samples of benthic invertebrates, 2nd edition. *Scientific Publications of the Freshwater Biological Association* No. 25, 160 pp.

Elliott, J. M. & Elliott, J. I. (1967). The structure and possible function of the buccopharyngeal teeth of *Sigara dorsalis* (Leech) (Hemiptera: Corixidae). *Proceedings of the Royal Entomological Society of London* (A), **42**, 83–86.

Elliott, J. M. & Tullett, P. A. (1978). A bibliography of samplers for benthic invertebrates. *Occasional Publications of the Freshwater Biological Association*, No. 4, 61 pp.

Elliott, J. M. & Tullett, P. A. (1983). A supplement to a bibliography of samplers for benthic invertebrates. *Occasional Publications of the Freshwater Biological Association*, No. 20, 26 pp.

Erlandsson, A., Malmqvist, B., Andersson, K. G., Herrmann, J. & Sjöström, P. (1988). Field observations on the activities of a group-living semiaquatic bug, *Velia caprai*. *Archiv für Hydrobiologie*, **112**, 411–19.

Fairbairn, D. J. (1986). Does alary polymorphism imply dispersal dimorphism in the waterstrider, *Gerris remigis? Ecological Entomology*, **11**, 355–68.

Fairbairn, D. J. (1988). Adaptive significance of wing dimorphism in the absence of dispersal: a comparative study of wing morphs in the waterstrider, *Gerris remigis*. *Ecological Entomology*, **13**, 273–81.

Fairbairn, D. J. & Desranleau, L. (1987). Flight threshold, wing muscle histolysis, and alary polymorphism: correlation traits for dispersal tendency in the Gerridae. *Ecological Entomology*, **12**, 13–24

Giller, P. S. (1980). The control of handling time and its effects on the foraging strategy of a heteropteran predator, *Notonecta*. *Journal of Animal Ecology*, **49**, 699–712.

Giller, P. S. (1986). The natural diet of the Notonectidae: field trials using electrophoresis. *Ecological Entomology*, **11**, 163–172.

Giller, P. S. & McNeill, S. (1981). Predation strategies, resource partitioning and habitat selection in *Notonecta* (Hemiptera/Heteroptera). *Journal of Animal Ecology*, **50**, 789–808.

Gillespie, E. (1985). Aquatic Hemiptera-Heteroptera of still waters from the Lothians, Scotland. *Entomologist's Monthly Magazine*, **121**, 125–6.

Groves, E. W. (1982). Hemiptera-Heteroptera of the London area, Part XII. *London Naturalist*, **61**, 72–87; 254–69.

Henrikson, L. & Oscarson, H. (1978a). A quantitative sampler for air-breathing aquatic insects. *Freshwater Biology*, **8**, 73–77.

Henrikson, L. & Oscarson, H. (1978b). Fish predation limiting abundance and distribution of *Glaenocorisa p. propinqua. Oikos*, **31**, 102–5.

Henrikson, L. & Oscarson, H. (1981). Corixids (Hemiptera-Heteroptera), the new top predators in acidified lakes. *Verhandlungen der internationalen Vereinigung für theoretische und angewandte Limnologie*, **21**, 1616–20.

Henrikson, L. & Oscarson, H. (1985). Waterbugs (Corixidae, Hemiptera-Heteroptera) in acidified lakes: Habitat selection and adaptations. *Ecological Bulletins*, **37**, 232–8.

Höregott, H. & Jordan, K. H. C. (1954). Bestimmungstabelle der Weibchen deutscher Corixiden (Heteroptera: Corixidae). *Beiträge zur Entomologie*, **4**, 578–94.

Jansson, A. (1969). Identification of larval Corixidae of Northern Europe. *Annales Zoologici Fennici*, **6**, 289–312.

Jansson, A. (1972). Mechanisms of sound production and morphology of the stridulatory apparatus in the genus *Cenocorixa* (Hemiptera, Corixidae). *Annales Zoologici Fennici*, **9**, 120–9.

Jansson, A. (1973). Stridulation and its significance in the genus *Cenocorixa* (Hemiptera, Corixidae). *Behaviour*, **46**, 1–36.

Jansson, A. (1976). Audiospectrographic analysis of stridulatory signals of some North American Corixidae (Hemiptera). *Annales Zoologici Fennici*, **13**, 48–62.

Jansson, A. (1977a). Distribution of Micronectae (Heteroptera, Corixidae) in Lake Päijänne, central Finland: Correlation with eutrophication and pollution. *Annales Zoologici Fennici*, **14**, 105–117.

Jansson, A. (1977b). Micronectae (Heteroptera, Corixidae) as indicators of water quality in two lakes in southern Finland. *Annales Zoologici Fennici*, **14**, 118–24.

Jansson, A. (1979a). Geographic variation in the stridulatory signals of *Arctocorisa carinata* (C. Sahlberg) (Heteroptera, Corixidae). *Annales Zoologici Fennici*, **16**, 36–43.

Jansson, A. (1979b). Reproductive isolation and experimental hybridisation between *Arctocorisa carinata* and *A. germari* (Heteroptera, Corixidae). *Annales Zoologici Fennici*, **16**, 89–104.

Jansson, A. (1979c). Experimental hybridisation of *Sigara striata* and *S. dorsalis* (Heteroptera, Corixidae). *Annales Zoologici Fennici*, **16**, 105–14.

Jansson, A. (1980). Post glacial distributional history of the water boatman, *Arctocorisa carinata* (Heteroptera, Corixidae). *Entomologia Generalis*, **6**, 235–45.

Jansson, A. (1981). A new European species and notes on synonymy in the genus *Corixa* Geoffroy (Heteroptera, Corixidae). *Annales Entomologici Fennici*, **47**, 65–68.

Jansson, A. (1986). The Corixidae (Heteroptera) of Europe and some adjacent regions. *Acta Entomologica Fennica*, **47**, 1–94.

Jansson, A. (1987). Micronectinae (Heteroptera, Corixidae) as indicators of water quality in Lake Vesijärvi, southern Finland, during the period 1976–1986. *Biological Research reports of the University of Jyväskylä*, **10**, 119–28.

Jansson, A. & Pajunen, I. (1978). Morphometric comparison of geographically isolated populations of *Arctocorisa carinata* (C. Sahlberg) (Heteroptera, Corixidae). *Annales Zoologici Fennici*, **15**, 132–42.

Jansson, A. & Scudder, G. G. E. (1972). Corixidae (Hemiptera) as predators: rearing on frozen brine shrimps. *Journal of the Entomological Society of British Columbia*, **69**, 44–45.

Jansson, A. & Scudder, G. G. E. (1974). The life cycle and sexual development of *Cenocorixa* species (Hemiptera: Corixidae) in the Pacific Northwest of North America. *Freshwater Biology*, **4**, 73–92.

Jastrey, J. T. (1981). Distribution and ecology of Norwegian water-bugs (Hem., Heteroptera). *Fauna Norvegica, Series B,*, **28**, 1–24.

Kaitala, A. (1987). Dynamic life history strategy of the waterstrider *Gerris thoracicus* as an adaption to food and habitat variation. *Oikos*, **48**, 125–31.

Kanyukova, E. V. (1986). [*Microvelia umbricola* Wrobl. – synonym of *M. buenoi* Drake (Heteroptera, Veliidae). In: Systematics and Ecology of Insects of the Far East.] The original is in Russian: *Sistematika i ec(k)ologia nasekomyth Dal'nego Vostoka* (ed. P. A. Lehr & A. N. Kupyanskaya), No. **13**. Institute of Biology and Pedology, Far Eastern Scientific Centre of the USSR Academy of Sciences, Vladivostok, USSR.

Kerfoot, W. C. & Sih, A. (1987). *Predation: Direct and indirect impacts on aquatic communities.* University Press of New England, Hanover, New Hampshire & London.

Kirby, P. (1983a). Heteroptera recorded from the Killarney area, 28 August – 7 September, 1981. *Irish Naturalists' Journal*, **21**, 45–7.

Kirby, P. (1983b). Corixidae from the South Lincolnshire Fens. *Entomologist's Monthly Magazine*, **119**, 43–9.

Kloet, G. S. & Hinks, W. D. (1964). *A check list of British Insects.* Part 1. Small Orders and Hemiptera. 2nd edition (revised); *Handbooks for the Identification of British Insects*, **11**, 1–119. Royal Entomological Society of London, London.

Lansbury, I. (1954). Some notes on the ecology of *Corixa (Halicorixa) stagnalis* Leach, with some information on measuring the salinity of brackish habitats. *Entomologist's Monthly Magazine*, **90**, 139–40.

Lansbury, I. (1965). Notes on the Hemiptera, Coleoptera, Diptera and other invertebrates of the Burren, Co. Clare and Inishmore, Aran Islands. *Proceedings of the Royal Irish Academy*, B, **64**, 89–115.

Lansbury, I & Leston, D. (1966). The distribution of *Sigara striata* in Britain. *Entomologist's Monthly Magazine*, **101**, 161–2.

Leston, D. (1955). Miscellaneous biological notes on British Corixidae and Notonectidae. *Entomologist's Monthly Magazine*, **91**, 92–5.

Leston, D. (1958). The distribution of water bugs (Hemiptera-Heteroptera: Hydrocorisae) in Ireland. *Entomologist's Monthly Magazine*, **94**, 26–31.

Macan, T. T. (1938). Evolution of aquatic habitats with special reference to the distribution of Corixidae. *Journal of Animal Ecology*, **7**, 1–19.

Macan, T. T. (1939). A key to the British species of Corixidae (Hemiptera-Heteroptera) with notes on their distribution. *Scientific Publications of the Freshwater Biological Association*, No. 1, 27 pp.

Macan, T. T. (1941). A key to the British water bugs (Hemiptera-Heteroptera excluding Corixidae). *Scientific Publications of the Freshwater Biological Association*, No. 4, 36 pp.

Macan, T. T. (1949). Corixidae (Hemiptera) of an evolved lake in the English Lake District. *Hydrobiologia*, **2**, 1–23.

Macan, T. T. (1954a). A contribution to the study of the ecology of Corixidae (Hemipt.). *Journal of Animal Ecology*, **23**, 115–41.

Macan, T. T. (1954b). The Corixidae (Hemipt.) of some Danish lakes. *Hydrobiologia*, **6**, 44–69.

Macan, T. T. (1955a). Littoral fauna and lake types. *Verhandlungen der Internationalen Vereinigung für theoretische und angewandte Limnologie*, **12**, 608–12.

Macan, T. T. (1955b). *Corixa lacustris* Macan a synonym of *C. dorsalis* Leach. *Hydrobiologia*, **7**, 124.

Macan, T. T. (1956). A revised key to the British water bugs (Hemiptera-Heteroptera). *Scientific Publications of the Freshwater Biological Association*, No. 16, 74 pp.

Macan, T. T. (1957). Some records of water bugs in Scotland. *Entomologist's Gazette*, **8**, 236–41.

Macan, T. T. (1965a). A revised key to the British water bugs (Hemiptera-Heteroptera). 2nd edition. *Scientific Publications of the Freshwater Biological Association*, no. 16, 78 pp.

Macan, T. T. (1965b). The fauna in the vegetation of a moorland fishpond. *Archiv für Hydrobiologie*, **61**, 273–310.

Macan, T. T. (1965c). Predation as a factor in the ecology of water bugs. *Journal of Animal Ecology*, **34**, 691–8.

Macan, T. T. (1967). The Corixidae of two Shropshire meres. *Field Studies*, **2**, 533–5.

Macan, T. T. (1970). *Biological Studies of the English Lakes*. Longman, London.

Macan, T. T. (1976). A twenty-one year study of the water bugs in a moorland fish pond. *Journal of Animal Ecology*, **45**, 913–22.

Macan, T. T. & Leston, D. (1978). *Notonecta striata* Linnaeus, 1758: Designation of a neotype under plenary powers. *Bulletin of Zoological Nomenclature*, **35**, 111–14. (followed by Opinion 1274. *Bulletin of Zoological Nomenclature*, **41**, 32–33 (1984).).

Macan, T. T. & Macfadyen, A. (1941). The water bugs of dew ponds. *Journal of Animal Ecology*, **10**, 175–83.

McCarthy, T. K. & Walton, G. A. (1980). *Sigara selecta* (Fieber) (Hemiptera/Heteroptera: Corixidae) new to Ireland, with notes on water bugs recorded from the Dingle Peninsula. *Irish Naturalists' Journal*, **20**, 64–6.

Maitland, P. S. (1977). *A coded checklist of animals occurring in fresh water in the British Isles*. Institute of Terrestrial Ecology, Edinburgh.

Martin, N. A. (1970). The distribution and ecology of the Corixidae (Hemiptera-Heteroptera) in Leicestershire. *Transactions of the Leicester Literary and Philosophical Society*, **64**, 101–22.

Morris, M. G. (1969). Associations of aquatic Heteroptera at Woodwalton Fen, Huntingdonshire, and their use in characterising artificial biotopes. *Journal of Applied Ecology*, **6**, 359–73.

Murray-Bligh, J. A. D. (1988). The macropterous form of *Aphelocheirus aestivalis* (F.) in Britain (Hemiptera: Aphelocheiridae). *Entomologist's Gazette*, **39**, 85–7

Nau, B. S. (1979). The relative frequency of Hertfordshire aquatic Hemiptera-Heteroptera. *Entomologist's Monthly Magazine*, **114**, 163–5.

Nieser, N. (1978). Heteroptera. *Limnofauna Europaea* (2nd edition) (ed. J. Illies), 280–5. Fischer, Stuttgart.

Nummelin, M., Vepsäläinen, K. & Spence, J. R. (1984). Habitat partitioning among developmental stages of waterstriders (Hemiptera: Gerridae). *Oikos*, **42**, 267–75.

O'Connor, J. P. (1986). Notes on *Scolopostethus puberulus* Horvath and *Limnoporus rufoscutellatus* (Latreille) in Ireland. *Entomologist's Record*, **98**, 33–5.

Oscarson, H. G. (1987). Habitat segregation in a water boatman (Corixidae) assemblage – the role of predation. *Oikos*, **49**, 133–40.

Pajunen, V. I. (1970). Phenology of the arrest of ovarian maturation in rock pool corixids (Heteroptera, Corixidae). *Annales Zoologici Fennici*, **7**, 270–2.

Pajunen, V. I. (1977). Population structure in rock pool corixids (Hemiptera, Corixidae) during the reproductive season. *Annales Zoologici Fennici*, **14**, 26–47.

Pajunen, V. I. (1979a). Competition between rock pool corixids. *Annales Zoologici Fennici*, **16**, 138–43.

Pajunen, V. I. (1979b). Quantitative analysis of competition between *Arctocorisa carinata* (Sahlb.) and *Callicorixa producta* (Reut.) (Hemiptera, Corixidae). *Annales Zoologici Fennici*, **16**, 195–200.

Pajunen, V. I. (1981). Analysis of developmental rates in field populations of *Arctocorisa carinata* (Sahlb.) and *Callicorixa producta* (Reut.) (Hemiptera, Corixidae) with the aid of developmental time axes. *Annales Zoologici Fennici*, **18**, 191–7.

Pajunen, V. I. (1982). Replacement analysis of non-equilibrium competition between rock pool corixids (Hemiptera, Corixidae). *Oecologia*, **52**, 153–5.

Pajunen, V. I. (1983). The use of physiological time in the analysis of insect stage frequency data. *Oikos*, **40**, 161–5.

Pajunen, V. I. & Sundbäck, E. (1973). Effect of temperature on the development of *Arctocorisa carinata* (Sahlb.) and *Callicorixa producta* (Reut.) (Hemiptera, Corixidae). *Annales Zoologici Fennici*, **10**, 372–7.

Pajunen, V. I. & Ukkonen, M. (1987). Intra- and interspecific predation in rock-pool corixids (Hemiptera, Corixidae). *Annales Zoologici Fennici*, **24**, 295–304.

Pearce, E. J. & Walton, G. A. (1939). A contribution towards an ecological survey of the aquatic and semi-aquatic Hemiptera-Heteroptera (water bugs) of the British Isles. *Transactions of the Society for British Entomology*, **6**, 149–80.

Peters, W. & Spurgeon, J. (1971). Biology of the waterboatman *Krisoosacorixa femorata* (Heteroptera: Corixidae). *American Midland Naturalist*, **86**, 197–207.

Poisson, R. (1922). Armature génitale et structure chitineuse du pénis dans le genre *Gerris* (Hem. Hydrometridae). *Bulletin de la Société entomologique de France*, **26**, 171–3.

Poisson, R. (1933). Les éspèces françaises du genre *Notonecta* L. et leurs principales formes affines paléarctiques. *Annales de la Société entomologique de France*, **102**, 317–58.

Popham, E. J. (1943). Ecological studies of the commoner species of British Corixidae. *Journal of Animal Ecology*, **7**, 124–36.

Popham, E. J. (1949). A contribution towards an ecological survey of the aquatic and semi-aquatic Hemiptera-Heteroptera (water bugs) of the British Isles. The Ribble Valley (Lancashire South and Mid). *Transactions of the Society for British Entomology*, **10**, 1–44.

Popham, E. J. (1950). Water bugs (Hemiptera-Heteroptera) of North Surrey. *Journal of the Society for British Entomology*, **3**, 158–73.

Popham, E. J. (1960). On the respiration of aquatic Hemiptera Heteroptera with special reference to the Corixidae. *Proceedings of the Zoological Society of London*, **135**, 209–42.

Popham, E. J. (1964). The migration of aquatic bugs with special reference to the Corixidae (Hemiptera-Heteroptera). *Archiv für Hydrobiologie*, **60**, 450–96.

Popham, E. J. (1966). An ecological study on the predatory action of the Three Spined Stickleback (*Gasterosteus aculeatus* L.). *Archiv für Hydrobiologie*, **62**, 70–81.

Popham, E. J., Bryant, M. T. & Savage, A. A. (1984). The role of front legs of British corixid bugs in feeding and mating. *Journal of Natural History*, **18**, 445–64.

Read, R. W. J. (1987). Records of local and uncommon Corixidae (Hemiptera) from West Cumbria. *Entomologist's Gazette*, **38**, 196.

Reynolds, J. D. (1975). Feeding in corixids (Heteroptera) of small alkaline lakes in Central B.C. *Verhandlungen der Internationalen Vereinigung für theoretische und angewandte Limnologie*, **19**, 3073–8.

Reynolds, J. D. & Scudder, G. G. E. (1987a). Experimental evidence of the fundamental feeding niche in *Cenocorixa* (Hemiptera: Corixidae). *Canadian Journal of Zoology*, **65**, 967–73.

Reynolds, J. D. & Scudder, G. G. E. (1987b). Serological evidence of realised feeding niche in *Cenocorixa* species (Hemiptera: Corixidae) in sympatry and allopatry. *Canadian Journal of Zoology*, **65**, 974–80.

Savage, A. A. (1971a). Some observations on the annual cycle of *Sigara concinna*. *Entomologist*, **104**, 230–2.

Savage, A. A. (1971b). *Sigara concinna* (Fieb.) (Hemiptera-Heteroptera) and Dyar's Law. *Entomologist*, **104**, 282–3.

Savage, A. A. (1971c). The Corixidae of some inland saline lakes in Cheshire, England. *Entomologist*, **104**, 331–44.

Savage, A. A. (1977). The effects of some environmental factors on the ionic concentration of an inland saline lake. *North Staffordshire Journal of Field Studies*, **17**, 1–11.

Savage, A. A. (1979a). The Corixidae of an inland saline lake from 1970 to 1975. *Archiv für Hydrobiologie*, **86**, 355–70.

Savage, A. A. (1979b). Some observations on oviposition in *Sigara concinna* (Fieber), *Sigara lateralis* (Leach) and *Sigara stagnalis* (Leach) (Hemiptera; Corixidae). *Archiv für Hydrobiologie*, **86**, 445–52.

Savage, A. A. (1981). The Gammaridae and Corixidae of an inland saline lake from 1975 to 1978. *Hydrobiologia*, **76**, 33–44.

Savage, A. A. (1982a). The survival and growth of *Gammarus tigrinus* Sexton (Crustacea: Amphipoda) in relation to salinity and temperature. *Hydrobiologia*, **94**, 201–12.

Savage, A. A. (1982b). Use of water boatmen (Corixidae) in the classification of lakes. *Biological Conservation*, **23**, 55–70.

Savage, A. A. (1985). The biology and management of an inland saline lake. *Biological Conservation*, **31**, 107–23.

Savage, A. A. (in preparation). The distribution of Corixidae in lakes and the ecological status of the North West Midlands meres with a key for the identification of species. *Field Studies*.

Savage, A. A. & Pratt, M. M. (1976). Corixidae (water boatmen) of the Northwest Midland meres. *Field Studies*, **4**, 465–76.

Scudder, G. G. E. (1976). Water-boatmen of saline waters (Hemiptera: Corixidae). *Marine Insects* (ed. L. Cheng), 263–289. North Holland Publishing Company, Amsterdam.

Scudder, G. G. E. (1983). A review of factors governing the distribution of two closely related corixids in the saline lakes of British Columbia. *Hydrobiologia*, **105**, 143–54.

Southwood, T. R. E. & Leston, D. (1959). *Land and water bugs of the British Isles*. Warne, London.

Spence, J. R. (1981). Experimental analysis of microhabitat selection in water striders (Heteroptera: Gerridae). *Ecology*, **62**, 1505–14.

Spence, J. R. (1983). Pattern and process in co-existence of water striders (Heteroptera: Gerridae). *Journal of Animal Ecology*, **52**, 497–511.

Spence, J. R. (1986). Relative impacts of mortality factors in field populations of the water strider *Gerris buenoi* Kirkaldy (Heteroptera: Gerridae). *Oecologia*, **70**, 68–76.

Spence, J. R., Spence, D. H. & Scudder, G. G. E. (1980). The effects of temperature on growth and development of water strider species (Heteroptera: Gerridae) of central British Columbia and implications for species packing. *Canadian Journal of Zoology*, **58**, 1813–20.

Spence, J. R. & Wilcox, R. S. (1986). The mating system of two hybridizing species of water striders (Gerridae) II. Alternative tactics of males and females. *Behavioural Ecology and Sociobiology*, **19**, 87–95.

Streams, F. A. (1987). Within habitat spatial separation of two *Notonecta* species: interactive vs. noninteractive resource partitioning. *Ecology*, **68**, 935–45.

Štys, P. & Jansson, A. (1988). Check-list of recent family-group and genus-group names of Nepomorpha (Heteroptera) of the world. *Acta Entomologica Fennica*, **50**, 1–44.

Sutton, M. F. (1947). The life history of *Corixa panzeri*. *Proceedings of the Linnaean Society of London*, 158, 51–62.

Sutton, M. F. (1951). On the food, feeding mechanism and alimentary canal of Corixidae (Hemiptera, Heteroptera). *Proceedings of the Zoological Society of London*, 121, 465–99.

Venkatesan, P. & Cloarec, A. (1988). Density dependent prey selection in *Ilyocoris* (Naucoridae). *Aquatic Insects*, 10, 105–16.

Vepsäläinen, K. (1971a). The roles of photoperiodism and genetic switch in alary polymorphism in *Gerris* (Het., Gerridae) (a preliminary report). *Acta Entomologica Fennica*, 28, 101–2.

Vepsäläinen, K. (1971b). The role of gradually changing daylength in determination of wing length, alary dimorphism and diapause in a *Gerris odontogaster* (Zett.) population (Gerridae, Heteroptera) in South Finland. *Annales Academicae Scientorum Fennicae A. IV. Biologica*, 183, 1–25.

Vepsäläinen, K. (1973). The distribution and habitats of *Gerris* Fabr. species (Heteroptera, Gerridae) in Finland. *Annales Zoologici Fennici*, 10, 419–44.

Vepsäläinen, K. (1974). The life cycles and wing lengths of Finnish *Gerris* Fabr. species (Heteroptera, Gerridae). *Acta Zoologica Fennica*, 141, 1–69.

Vepsäläinen, K., Kaitala, A. & Kaitala, V. (1985). Reproductive tactics of the water strider *Gerris thoracicus* in unpredictable environments: a simulation study. *Oikos*, 45, 266–72.

Vepsäläinen, K. & Krajewski, S. (1986). Identification of the waterstrider (Gerridae) nymphs of Northern Europe. *Annales Entomologici Fennici*, 52, 63–77.

Vepsäläinen, K. & Nummelin, M. (1985a). Female territoriality in the waterstriders *Gerris najas* and *G. cinereus*. *Annales Zoologici Fennici*, 22, 433–9.

Vepsäläinen, K. & Nummelin, M. (1985b). Male territoriality in the water strider *Limnoporus rufoscutellatus*. *Annales Zoologici Fennici*, 22, 441–8.

Vepsäläinen, K. & Nummelin, M. (1986). Habitat selection by water strider larvae (Heteroptera: Gerridae) in relation to food and imagoes. *Oikos*, 47, 374–81.

Walton, G. A. (1936). Oviposition in the British species of *Notonecta* (Hemipt.). *Transactions of the Society for British Entomology*, 3, 49–57.

Walton, G. A. (1942). The aquatic Hemiptera of the Hebrides. *Transactions of the Royal Entomological Society of London*, 92, 417–52.

Walton, G. A. (1943). The water bugs (Rhyncota-Hemiptera) of North Somerset. *Transactions of the Society for British Entomology*, 8, 231–90.

Walton, G. A. (1981). *Microvelia pygmaea* (Dufour, 1833) (Hemiptera: Veliidae), a flightless water bug new to Ireland. *Irish Naturalists' Journal*, 20, 223–8.

Wilcox, R. S. & Spence, J. R. (1986). The mating system of two hybridizing species of water striders (Gerridae) I. Ripple signal functions. *Behavioural Ecology and Sociobiology*, 19, 79–85.

Wróblewski, A. (1938). Eine neue Veliiden-Art aus Polen (Heteroptera). Novy gatunek krajowy z rodziny Veliidae (Heteroptera). *Annales Musei Zoologici Polonici*, 13, 213–18.

Young, E. C. (1965a). The incidence of flight polymorphism in British Corixidae and descriptions of the morphs. *Journal of Zoology*, 146, 567–76.

Young, E. C. (1965b). Flight muscle polymorphism in British Corixidae: ecological observations. *Journal of Animal Ecology*, **34**, 353–90.

Zera, A. J. (1985). Wing polymorphism in waterstriders (Gerridae: Hemiptera): mechanisms of morph determination and fitness differences between morphs. *Migration: mechanism and adaptative significance* (ed. M. A. Rankin & H. Dingle), 647–86. *Contributions to Marine Sciences of the University of Texas*, 27 (supplement).

INDEX

The index includes all taxa appearing in the text. Synonyms, genera and subgenera rejected are shown in parentheses. Page numbers in **bold type** indicate main key references.

PUBLICATIONS OF THE FRESHWATER BIOLOGICAL ASSOCIATION

These publications and an up to date price list may be obtained direct from **Dept. DWS, Freshwater Biological Association, The Ferry House, Ambleside, Cumbria, LA22 0LP.**

SCIENTIFIC PUBLICATIONS

5. A KEY TO THE BRITISH SPECIES OF FRESHWATER CLADOCERA, by the late D. J. Scourfield and J. P. Harding, 3rd ed., 1966.   ISBN 0 900386 01 0

13. A KEY TO THE BRITISH FRESH- AND BRACKISH-WATER GASTROPODS, by T. T. Macan, 4th ed., 1977.   ISBN 0 900386 30 4

17. A KEY TO THE ADULTS AND NYMPHS OF THE BRITISH STONEFLIES (PLECOPTERA), by H. B. N. Hynes, 3rd ed., 1977. (Reprinted 1984).   ISBN 0 900386 28 2

18. A KEY TO THE BRITISH FRESHWATER CYCLOPID AND CALANOID COPEPODS, by J. P. Harding and W. A. Smith, 2nd ed., 1974.   ISBN 0 900386 20 7

23. A KEY TO THE BRITISH SPECIES OF FRESHWATER TRICLADS, by T. B. Reynoldson, 2nd ed., 1978.   ISBN 0 900386 34 7

25. SOME METHODS FOR THE STATISTICAL ANALYSIS OF SAMPLES OF BENTHIC INVERTEBRATES, by J. M. Elliott, 2nd ed., 1977.   ISBN 0 900386 29 0

27. A KEY TO BRITISH FRESHWATER FISHES, by Peter S. Maitland, 1972.
     ISBN 0 900386 18 5

29. TURBULENCE IN LAKES AND RIVERS, by I. R. Smith, 1975. ISBN 0 900386 21 5

30. AN ILLUSTRATED GUIDE TO AQUATIC AND WATER-BORNE HYPHOMYCETES (FUNGI IMPERFECTI), by C. T. Ingold, 1975.   ISBN 0 900386 22 3

31. A KEY TO THE LARVAE, PUPAE AND ADULTS OF THE BRITISH DIXIDAE (DIPTERA), by R. H. L. Disney, 1975.   ISBN 0 900386 23 1

32. A KEY TO BRITISH FRESHWATER CRUSTACEA MALACOSTRACA, by T. Gledhill, D. W. Sutcliffe and W. D. Williams, 1976.   ISBN 0 900386 24 X

33. DEPTH CHARTS OF THE CUMBRIAN LAKES, by A. E. Ramsbottom, 1976.
     ISBN 0 900386 25 8

34. AN ILLUSTRATED KEY TO FRESHWATER AND SOIL AMOEBAE, by F. C. Page, 1976.
     ISBN 0 900386 26 6

35. A KEY TO THE LARVAE AND ADULTS OF BRITISH FRESHWATER MEGALOPTERA AND NEUROPTERA, by J. M. Elliott, 1977.   ISBN 0 900386 27 4

36. WATER ANALYSIS: SOME REVISED METHODS FOR LIMNOLOGISTS, by F. J. H. Mackereth, J. Heron & J. F. Talling, 1978
(Second Impression, 1989).   ISBN 0 900386 31 2

37. A KEY TO THE ADULT MALES OF THE BRITISH CHIRONOMIDAE (DIPTERA), by L. C. V. Pinder, 1978.   ISBN 0 900386 32 0

38. A KEY TO THE FRESHWATER PLANKTONIC AND SEMI-PLANKTONIC ROTIFERA OF THE BRITISH ISLES, by Rosalind M. Pontin, 1978.   ISBN 0 900386 33 9

39. A Guide to Methods for Estimating Microbial Numbers and Biomass in Fresh Water, by J. Gwynfryn Jones, 1979. ISBN 0 900386 37 1

40. A Key to the British Freshwater Leeches, by J. M. Elliott and K. H. Mann, 1979. ISBN 0 900386 38 X

41. A Key to the British and European Freshwater Bryozoans, by S. P. Mundy, 1980. ISBN 0 900386 39 8

42. Desmids of the English Lake District, by Edna M. Lind & Alan J. Brook, 1980. ISBN 0 900386 40 1

43. Caseless Caddis Larvae of the British Isles, by J. M. Edington & A. G. Hildrew, 1981. ISBN 0 900386 41 X

44. A Guide to the Morphology of the Diatom Frustule with a key to the British Freshwater Genera, by H. G. Barber and E. Y. Haworth, 1981. ISBN 0 900386 42 8

45. A Key to the Larvae of the British Orthocladiinae (Chironomidae), by P. S. Cranston, 1982. ISBN 0 900386 43 6

46. The Parasitic Copepoda and Branchiura of British Freshwater Fishes: a Handbook and Key, by Geoffrey Fryer, 1982. ISBN 0 900386 44 4

47. A Key to the Adults of the British Ephemeroptera, by J. M. Elliott & U. H. Humpesch, 1983. ISBN 0 900386 45 2

48. Keys to the Adults, Male Hypopygia, Fourth-instar Larvae and Pupae of the British Mosquitoes (Culicidae), by P. S. Cranston, C. D. Ramsdale, K. R. Snow and G. B. White, 1987. ISBN 0 900386 46 0

49. Larvae of the British Ephemeroptera: A Key with Ecological Notes, by J. M. Elliott, U. H. Humpesch & T. T. Macan, 1988. ISBN 0 900386 47 9

OCCASIONAL PUBLICATIONS